Crisis and Emergency Management in the Arctic

This book sheds light on the management challenges of crisis and emergency response in an Arctic environment.

It explores how the complexity of the operational environment impacts on the risk of operations and addresses a need for tailor-made emergency response mechanisms. Through case studies of the Arctic environment, the book illustrates how factors such as nature, geography, demographics, and infrastructure increase the complexity of crises in the Arctic and present a significant danger to life and health, the environment, and values in challenging Arctic waters. The case studies place a special focus on contextual factors including conflicting interests and different stakeholder groups, as well as the institutional platforms influencing crisis response and emergency management. They also explore the implications for the managerial roles, the mode of operations, and the structuring of the organizations responsible for the emergency response. The necessity to facilitate co-operation across organizations and borders and a need for organizational flexibility in large-scale operations are also emphasized.

Written in an accessible style, this book will make for a useful resource for undergraduate and postgraduate students of disaster and emergency management, as well as for professionals involved in emergency services.

Natalia Andreassen is Associate Professor in Organization and Management at the Nord University Business School, Norway. Her research focuses on issues of co-ordination and organization, with a special focus on emergency management. She conducts research at the High North Center of Business and Governance.

Odd Jarl Borch is Professor of Strategy at the Nord University, Bodo, Norway. He received his Master's in Science degree. from The Norwegian School of Economics and Business Administration in 1979 and his PhD from Umea University in Sweden in 1990. He has a Master's degree in Mariner Education from Bodin Maritime Academy.

Routledge Studies in Hazards, Disaster Risk and Climate Change
Series Editor: Ilan Kelman, *Reader in Risk, Resilience and Global Health at the Institute for Risk and Disaster Reduction (IRDR) and the Institute for Global Health (IGH), University College London (UCL)*

This series provides a forum for original and vibrant research. It offers contributions from each of these communities as well as innovative titles that examine the links between hazards, disasters and climate change, to bring these schools of thought closer together. This series promotes interdisciplinary scholarly work that is empirically and theoretically informed, with titles reflecting the wealth of research being undertaken in these diverse and exciting fields.

For more information about this series, please visit: https://www.routledge.com/Routledge-Studies-in-Hazards-Disaster-Risk-and-Climate-Change/book-series/HDC

Crisis and Emergency Management in the Arctic

Navigating Complex Environments

Edited by
Natalia Andreassen and Odd Jarl Borch

Routledge
Taylor & Francis Group

LONDON AND NEW YORK

First published 2021
by Routledge
2 Park Square, Milton Park, Abingdon, Oxon OX14 4RN

and by Routledge
52 Vanderbilt Avenue, New York, NY 10017

Routledge is an imprint of the Taylor & Francis Group, an informa business

British Library Cataloguing in Publication Data
A catalogue record for this book is available from the British Library

Library of Congress Cataloging-in-Publication Data
A catalog record has been requested for this book

ISBN: 978-0-367-14055-7 (hbk)
ISBN: 978-0-429-02989-9 (ebk)

Typeset in Times New Roman
by Taylor & Francis Books

Contents

Illustrations

Figures

Tables

Preface

The aim of this book is to explore the managerial challenges in crises and emergency response in a complex environment, with a special focus on the Arctic context.

We take as a starting point the responsibility of companies, governments, and local communities to prepare for and build up a capacity to respond to incidents threating life, health, and the environment, as well as economic and social values. In this book we reflect upon crises that could represent a significant danger to life and health, the environment, and societal values, with a special focus on maritime risk areas. In all countries, a mix of government emergency responders, companies, organizations, and communities, as well as volunteers, are ready to face the challenge, reduce the consequences of dangerous incidents, and bring organizations and society back to normal after a crisis. In this book, we elaborate on the organization and management of the emergency response forces and discuss adaptations in emergency responses in the Arctic context.

We start with a focus on how the context such as nature, geography, demographics, and infrastructure, as well as politics, increases the complexity of the operations and influence on the risk picture. We reflect on maritime activity in challenging regions in the Arctic and the risk patterns. We address the difficulties in assessing risk and vulnerabilities and the role of institutional regulations and agreements in building a platform for risk reduction and emergency response.

Second, the book introduces us to emergency preparedness resources and the operational patterns of the preparedness agencies. The authors have in mind the potential for increased cross-institutional and cross-border co-operation and discuss avenues for further agreements and co-operative relations.

We thank our financial contributors – the Norwegian Ministry of Foreign Affairs, the Nordland County Administration, the Research Council of Norway, the High North Center for Business and Governance at Nord University Business School, and all our partners in the Maritime Preparedness and International Partnership in the High North (MARPART) and Inter-organizational Co-ordination of Mass Rescue Operations in Complex

Environments (MAREC) projects. We are grateful to the leaders of the emergency response agencies, companies in the Arctic countries, and members of the MARPART Advisory Board for their great support.

Bodø, February 2020
Natalia Andreassen and Odd Jarl Borch

Contributors

Natalia Andreassen is an Associate Professor of organization and management at Nord University Business School, Norway. She is a co-ordinator of the Master's Program in Preparedness and Emergency Management at Nord University Business School. She holds a PhD from Nord University Business School. Natalia is conducting research within the field of emergency management, with a special focus on maritime emergency preparedness, the organization and co-ordination of incident response management, risk assessments, and collaboration exercises. She has been working as a researcher at the High North Center of Business and Governance. She has been working with the MARPART projects, an international research and development consortium on maritime safety and security in the High North, and the MAREC project on inter-organizational co-ordination of mass rescue operations in complex environments.

Ole Kristian Bjerkemo was born in Norway and graduated from Army Officer's School and the Norwegian Business School (BI). He has more than 30 years of experience in emergency planning and response in positions in the Norwegian armed forces and Homeguard, Civil Defence, as a consultant, and in different positions in the Norwegian Pollution Control Authority and the Norwegian Coastal Administration. He has been chair for the Bonn Agreement working group and the Arctic Council working group for emergency prevention, preparedness, and response. In recent years he has been international co-ordinator in the Department of Emergency Response in the Norwegian Coastal Administration.

Odd Jarl Borch is a Professor of Strategy at the Nord University Business School, Bodø, Norway and a part-time professor in Emergency Management at the Arctic University of Norway (UiT). He received his Master's in Science degree. from the Norwegian School of Economics and Business Administration in Bergen in 1979 and his PhD from Umeå University in Sweden in 1990. He has a Master's degree in Mariner Education from Bodin Maritime Academy. He is conducting research within the field of organization and strategic management, focusing on entrepreneurship and innovation within the marine and maritime industries. He leads several projects on maritime safety, security, and emergency preparedness, with a special emphasis on the Arctic. He is the author of more than 150 scientific publications and has served on several government committees within industry development, safety, and preparedness. He is the founder and first leader of the University of the Arctic Thematic Network on Arctic Safety and Security, which includes 21 universities and research institutions in nine countries. He represented the UiT as head of the delegation of observers within the Arctic Council special committee on Emergency Prevention, Preparedness, and Response and Protection of the Arctic Marine Environment.

Dimitrios Dalaklis joined the World Maritime University as an Associate Professor of the Maritime Safety and Environmental Administration Specialization in 2014, following 26 years of distinguished service with the Greek Navy. He is a Naval Postgraduate School (USA) graduate, and his PhD research was conducted at the University of the Aegean (Greece, Department of Shipping, Trade, and Transport). His expertise revolves around the extended safety and security domains, including issues related to the conduct of navigation (regulatory framework, techniques in associated best practices, and related equipment). He is the author/co-author of many books articles and studies in Greek and English.

Lieutenant Commander Megan Drewniak joined the US Coast Guard in 2004, and she is currently serving as the Commanding Officer of US Coast Guard Marine Safety Unit Toledo, Ohio. In this capacity, she is responsible for executing the Coast Guard's Port Safety and Security, Marine Environmental Protection, and Commercial Vessel Safety missions from Monroe, Michigan to Huron, Ohio. Prior to this stationing, she was the Coast Guard Liaison Officer/Lecturer to the World Maritime University in Malmö, Sweden. She is the author/co-author of numerous articles and publications with a particular focus on maritime governance and freedom of navigation in the Arctic region.

Edda Falk is the Communications Manager of the Association of Arctic Expedition Cruise Operators (AECO). A Norwegian national, her background includes language studies, communications, and international relations, specializing in Arctic issues. She has previously worked for the Norwegian Embassy in Havana, Cuba, and at the Canadian Embassy in Oslo, Norway. She has worked as an arctic advocacy officer for the Canadian International Arctic Centre in Oslo, where she co-ordinated the promotion of Canada's Arctic interests abroad. She has also held the position of communications adviser at the High North Center for Business and Governance in Bodø, which focuses on business and innovation in the circumpolar Arctic. Since joining AECO in 2017, Edda has managed several projects, including the development of AECO's communication strategy and crisis management tools. She has also been involved in developing

AECO's Off-Vessel Risk Assessment Tool, a mobile app to be used in Arctic field operations.

Petr Gerasun is the Deputy Chief of the Maritime Rescue Service and the Head of the State Maritime Rescue Co-ordination Centre in Moscow, Russia. He graduated in 1978 from middle school in Murmansk, USSR, and after that completed two years' national service in the army. He continued his education at Engineering Maritime College from which he graduated in 1988 in as a Navigation Engineer. He received higher education in law from Murmansk State Technical University in 1997. During 1983–99 he worked his way up from a deck hand to a master on board fishing vessels, cargo ships, and tankers. In 1999 he started to work as a co-ordinator in the State Maritime Rescue Co-ordination Centre in Murmansk and from 2011 he served as the head of that body. In January 2020, he took charge of the State Maritime Rescue Co-ordination Centre in Moscow, Russia.

Jens Peter Holst-Andersen has been the Chair of the Arctic Council Working Group Emergency Prevention, Preparedness and Response (EPPR) since 2017. Prior to that, he was the Head of the Danish Delegation to the EPPR and on the founding team of the Arctic Coast Guard Forum. As a navy officer, his career has encompassed the Arctic region on the operational, strategic, and policy levels in the Danish armed forces and currently the Danish Ministry of Defence. He has had a leading role in projects related to environmental protection and search and rescue in the Arctic, including advisory roles on international projects.

Bent-Ove Jamtli is the Director of the Joint Rescue Coordination Centre (JRCC) North Norway and the ARCSAR project co-ordinator. He has been the director of JRCC North-Norway since 2013. He was previously a fighter jet controller in the Norwegian air force and has held several staff and command positions. He worked as a search and rescue mission co-ordinator at the JRCC in northern Norway in 1996–2000. He was then in charge of planning and execution of the international NATO/PfP search and rescue exercise 'Bright Eye'. He has a Master's degree in political science from Nord University, Bodø.

Frigg Jørgensen is the Executive Director of AECO and has been with the organization since 2006. Frigg comes from northern Norway and lived in Svalbard for more than 30 years before relocating to Tromsø, where AECO's head office is located. She has worked in polar tourism since 1992. After completing a college degree in tourism, she started her career as a manager for what became Visit Svalbard, building up the local tourist office in Longyearbyen and co-ordinating the establishment of the Svalbard Tourism Board. From 2001 until 2006 she worked as the first tourism adviser at the office of the Governor of Svalbard. Frigg has been the project manager for the development of "The Svalbard Guide Courses",

"General guidelines for tourism in Svalbard", "Clean up Svalbard", and "The Governor's strategy for tourism and outdoor recreation". Since joining AECO, she has been responsible for several of AECO's numerous projects, including establishing the Joint Arctic SAR TTX and Workshop.

Svetlana Kuznetsova works in the Arkhangelsk Regional Rescue Service located in Arkhangelsk, Russia. She has over ten years' experience in emergency prevention and response in northern Russia. Svetlana has worked on several international projects related to emergency preparedness and held a researcher position at the Northern Arctic Federal University within the international project "MARPART: Maritime preparedness and International Partnership in the High North". She has a Master's Degree in ecology, during which she examined risk management of oil spills in the Arctic. She is continuing her studies as a PhD student at the Northern Arctic Federal University.

Nataliya Marchenko graduated from Lomonosov Moscow State University in 1988 and worked there during 1992–2006 in the Faculty of Geography. During 1988–1991 Nataliya worked in the Far East branch of the Russian Academy of Science (Vladivostok) and did a PhD project on forest fire risk assessment, using a geographical information system (GIS). She defended her PhD thesis in Geography in 1992 at the Institute of Geography Russian Academy of Science (Moscow). Since 2006 she has worked at the University Centre in Svalbard, Norway in the Arctic Technology department and does research in sea ice, ice-induced accidents, shipping in the Arctic, and risk assessment. She is the author of more than 50 scientific publications. Her book *Russian Arctic Seas. Navigation Conditions and Accidents* was published by Springer in 2012. Her other field of research is GIS and Data Analysis for Arctic Technology. She is a co-ordinator of several scientific and collaboration projects, and she makes, designs, and updates websites and e-rooms. In the MARPART consortium, Nataliya is responsible for the Svalbard region, including the estimation of the activity levels, risk assessments, and mapping of marine emergencies (GIS online).

Tor-Geir Myhrer is Professor in Police Law at the Norwegian Police University College. Before joining the College, he had 23 years of experience as a prosecutor (of which 15 years at the Office of the Director of Public Prosecutions) and legal adviser in the law department of the Norwegian Ministry of Justice and the Police. He was a part-time professor in criminal law and criminal procedure law at the Faculty of Law, University of Oslo from 2005 to 2011. His area of expertise is criminal procedure, confidentiality obligation within the police and public prosecutor authorities, police power, and police and prosecution ethics.

Torbjørn Pedersen is a Professor of Political Science at Nord University, Norway. He holds a PhD degree in Political Science from the University of Tromsø and another Master's degree in International Relations and

International Economics while a Fulbright fellow at SAIS Johns Hopkins University, Washington, DC. He has specialized in international politics and Arctic affairs. He is a former journalist with Reuters, *Aftenposten*, and the Norwegian Broadcasting Corporation (NRK).

Rebecca Pincus is an Assistant Professor in the Strategic and Operational Research Department at the Center for Naval Warfare Studies at the US Naval War College (NWC). Within the NWC, she is an associate of the Russia Maritime Studies Institute and the Institute for Future Warfare Studies. Her areas of research lie at the nexus of national security and the natural environment, with an emphasis on climate change and resource-based topics. Her current research interests centre on the strategic implications of an opening Arctic, including for the North Atlantic/Greenland, Iceland, UK region and the Russian Arctic. In addition to the Arctic region, she maintains a research interest in Antarctica. Dr Pincus previously served as a primary investigator at the US Coast Guard's Center for Arctic Study and Policy, where she executed research-based projects for the Coast Guard on a range of polar topics, collaborating with a wide variety of the US Coast Guard's offices. She was a Fulbright Fellow in Iceland in 2015 with the Icelandic Ministry of Foreign Affairs, also working closely with the Icelandic Coast Guard and teaching at the University of Iceland. Her work has appeared in *Polar Journal, Polar Geography, War and Society, War on the Rocks*, and more. In 2015 Yale released *Diplomacy on Ice: Energy and the Environment in the Arctic and Antarctica*, which Dr Pincus co-edited with Dr Saleem Ali. Dr Pincus completed her PhD in 2013 at the University of Vermont's Rubenstein School of the Environment and Natural Resources, including a Master's degree in Science in 2010. She also holds a Master's degree in Environmental Law from the Vermont Law School (2011).

Jana Prochotska received her Master's degree in International Relations from the University of Economics in Bratislava, Slovakia. She is a PhD candidate at the Academy of the Police Force in Bratislava, and her PhD thesis is elaborated within emergency and preparedness management and public administration. In the course of her PhD studies, she spent eight months at the Police Academy of the Netherlands, where the main areas of study included terrorism, cyber-terrorism, and organized crime. She is a researcher at the Norwegian Police University College, and within the MARPART project she specializes in prevention and response to violent acts, terrorism, and cyber-attacks, mainly in relation to soft targets.

Ensieh Roud received her Master's degree in Maritime Management from University College in south-eastern Norway. She is a PhD candidate at Nord University; in parallel, she is working on the MARPART project as a researcher at the High North Center for Business and Governance. Her main research area is training and education within emergency management and inter-organizational collaboration.

Johannes Schmied holds a PhD scholar position at the High North Center for Business and Governance at Nord University Business School. He has also worked as an adviser for Nord University and has been the country manager for FSC-certified timber companies in Latin America. He has been working on several research and development projects on emergency management co-operation in the High North, including MARPART projects. His PhD research topic focuses on the development of crisis management capability under complexity. His concept of crisis involves unwanted and unexpected events, which cause threats to the existence of individuals, organizations, or systems. Crises can be related to safety and security, finances, natural disasters, among other things. He is interested in the involved organizations and their managers, as well as the environment and incidents that affect it, which are factors that influence complexity.

Captain Roberto H. Torres serves as the Chief of Regional Affairs in the Office of International Affairs and Foreign Policy at the US Coast Guard headquarters in Washington, DC. Captain Roberto Torres is US Coast Guard Representative to the Arctic Coast Guard Forum.

1 Introduction to Crisis and Emergency Management in the Arctic

Natalia Andreassen and Odd Jarl Borch

Introduction

The interdisciplinary research tradition of crisis and emergency management builds upon insights from other disciplines like risk management, political science, law, institutional theory, organizational studies, and management. Crisis and emergency management is a budding research field continuously developing, not the least driven by perceptions and inputs from practice. New types of crises and emergencies, especially 'black swan' events, challenge practitioners and call for new knowledge. We link the analyses in this book to some of the themes of discussion within the crisis and emergency management system of the Arctic, reflect upon complexity issues, and illuminate how the Arctic context could provide challenges for the lead emergency response organizations. In this book, we include reflections from experienced professionals shedding light on some of the operational challenges and areas for further developments in this field.

Managing larger crises in the remote areas of the Arctic can be a demanding task that requires specialized competencies and experiences to deal with the different roles and tasks. The central idea of the book is to illuminate a potential large-scale accident and the management of a combined emergency response operation from a complexity perspective. We focus on the lessons learnt from real cases and exercises and the need for more efficient institutional collaboration, resource capability development, and emergency system management.

A special emphasis is place on inter-organizational and cross-border co-operation and the organizational structure related to co-ordination and control within joint operations on large-scale incidents, which demand support from many organizations and often support from neighbouring countries. The emergency response systems in most countries are characterized by strict structures, a high degree of formalization, and a range of standard operating procedures for different kinds of response operations, including task lists and clear-cut functional descriptions. There is a command hierarchy, written communication routines, and a broad set of laws and regulations behind the operational system. The studies in this book add to the organizational literature that discusses how organizations might have to adapt their processes triggered by factors within and outside the organizations. The contributors argue that with increased complexity, emergency response managers might have to deviate from the established organizational structures and management principles to deal with a new context and new tasks.

In the first part of this book we discuss how the task environments could include more factors to survey, more links between the present and new factors, differences in activities, and a lack of understanding of cause–effect relations, leading to and affecting the consequences of an accident. Complexity is defined by the potentially incomplete cues that decision-makers need to base their judgements on (Tversky and Kahneman, 1974; Kahneman and Klein, 2007; Steigenberger et al., 2017). When there is a lack of statistics and experience, the emergency response management teams do not have certainty in the cues–outcomes relationship and will rely on their probability judgements. Newcomers with limited experience, the limitations of equipment in cold climate conditions, a challenging natural environment, and a lack of large-scale emergency management experiences in the Arctic region could affect the emergency response system. An exchange of expert judgements and assessments of the operational conditions, improved predictions of weather, and a focus on choosing the safest route could potentially reduce complexity for commercial and governmental activity in the Arctic and could have implications for risk reduction strategies. However, there may be 'rest factors' – the accidents that we cannot avoid, and the incidents that are difficult to predict.

Complexity is also related to the range of organizational levels with the increasing number of interdependencies of heterogeneous elements – teams of organizations, jurisdictions, and management levels (Czarniawska, 2004; Weick and Sutcliffe, 2011). Complexity in crisis and emergency operations could relate to responsive processes that connect people's interaction and behaviour with the change and unpredictability of social realities (Johannessen, 2018). Large-scale emergencies in the Arctic will include a range of actors of various jurisdictions and responsibilities and different levels of command and formalities within the institutional regimes. It is essential to be familiar with the different sub-systems and how they can be inter-linked. In specific situations there might be a need for instant, on-the-spot tailor-making in the forms of co-operation.

We also talk about task complexity. Task complexity is defined as the number of components and the ties between them that can provide alternative routes towards a goal (Campbell, 1988; Hærem et al., 2015). The concept extended by Hærem et al. (2015) emphasizes tasks performed by multiple actors at different levels. These tasks are regarded as dynamic processes that can change over time. They are connected to social and material contexts. Task complexity can be seen as an independent variable having an impact on organizational learning (Hærem et al., 2015), emergency co-ordination processes (Wolbers et al., 2017, Gephart et al., 2019), or emergency strategies (Johannessen, 2018). In the context of large-scale incidents in the Arctic, the increased task complexity can call for flexibility and improvisational management with greater freedom from pre-established procedures and strategies. This also has potential implications for competence development and training.

Figure 1.1 shows the main elements discussed in this book. The studies add to the literature within activity and risk aspects in the Arctic, the institutional platforms for risk reduction and emergency response, crisis and emergency management, and knowledge transfer.

The studies in this book are built on single-case and multiple-case analyses, documentary studies, and literature discourse. The chapters include primary data from interviews, conversations, and participant observations, as well as secondary data such as documents and agreements, incident reports, exercise reports, and internal documents obtained from the responders. An additional source of information are the expert commentaries from experienced professionals in this book.

Activity and Risk Aspects in the Arctic

Maritime activity in the North varies a lot between the different sea areas. Natural resources, industrial development, and other socioeconomic factors in the various countries have influenced the maritime infrastructure and maritime activity (Borch et al., 2016a). Maritime activity is at its highest level in the Norwegian sectors of

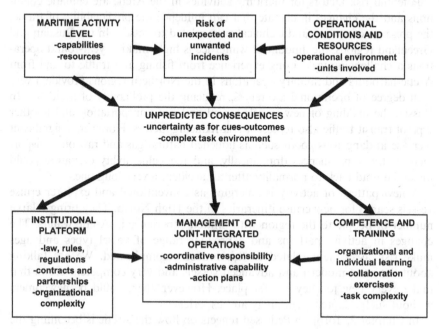

Figure 1.1 The management of integrated emergency response operations in the Arctic.

the Arctic, owing to the fisheries industry and increasing cruise traffic. As for intercontinental traffic, vessels through the Northeast Passage and the Northern Sea Route, passing through Russian coastal waters and along the coast of main-land Norway, have increased. The increasing numbers of tankers from Russia is leading to an increasing volume of intercontinental traffic through Norwegian Arctic waters as well. Fishing activity has always been the key industry in many Arctic states. This activity has seasonal and geographical differences. Some of the trawler fleets are operating very close to and sometimes into the ice ridge on a year-round basis. The oil and gas activity in the Arctic is a special case for Russia because it produces 11% of Russia's national income and around 22% of total Russian exports. Maritime tourism is a growing activity in the High Arctic. At least four types of cruise concepts are present within Arctic maritime tourism –overseas (conventional) cruises, expedition cruise vessels, day trip ships, and yachts.

Research activity has traditionally been important in the Arctic. The main research and other governmental activities in the Arctic include the search for natural resources, protecting borders and resources, naval activity, and polar research. Activity is increasing in the region, not least from navy vessels. There are also new transport routes to reopened military bases in the North.

The risk perspective is an important starting point for the discussion on emergency management and preparedness capacities. As for maritime activity and risk patterns in the High North, the focus has been on the dominating risk factors, risk types, and the probability of unwanted incidents in the Arctic region (Borch et al., 2016b).

Potential risk factors for maritime activities in the Arctic are climatic condi-tions and the changes in climate that can influence the predictability of traffic, the presence of sea ice, and technical risks related to vessels. In the Russian and Greenland regions, the amount of winter ice is high, making year-round opera-tions a challenging task. Long experience from fishing and traffic to and from Arctic harbours and military operations in the Northern regions provide a very high degree of operational awareness, reducing the probability of accidents. In Russia, the building of new vessels for oil and gas transportation and another type of transit traffic also represent risk-reducing factors. However, a significant increase in dangerous goods such as liquefied natural gas and raw oil transport increase the consequences dramatically, and the vulnerability of nature could make the road back to normality after an accident a very long one.

A new pattern of activity is emerging as conventional and explorer cruise vessels search for new cruise itineraries in the High North. They bring a large number of people to the region with limited experience in Arctic waters. The changes in activity patterns and a broader range of vessel types and ages make risk assessment related to this industry complicated. We are talking about the most modern and advanced vessels and very competent crews that make every new journey a safer place. However, there could be newcomers without such adapted technology and experiences.

In Chapter 2, Torbjørn Pedersen reflects on how the Arctic is becoming the arena for new countries with a political, economic, and/or military interests in the Arctic. He shows how the Arctic has long been an exclusive operational area for Arctic countries, but is becoming more inclusive, with newcomers from a large part of Europe, as well as Asian countries. From a situation awareness perspective, he illuminates how polar research, international research co-operation, and the sharing of environmental data enable far more operators to focus on the Arctic. He emphasizes how the higher levels of overall situational awareness open the Arctic for new stakeholders with potential conflicting interests.

In Chapter 3, Nataly Marchenko describes the challenges of accurate risk assessment in the Arctic. She reflects on the risk patterns for different types of accidents and in different regions of the Atlantic Arctic. Various theories and perspectives on risk assessment are discussed. As marine accidents in the Arctic are rare, there are limited statistics available. Therefore, qualitative analysis and expert judgements provide the basis for many risk assessments. She suggests that for emergency response development, an assessment of the risk on the regional level, using a Regional Risk Index calculation, should be developed. Furthermore, for the emergency response system, looking into unlikely events, the analysis of worst-case scenarios such as 'black swan' events, could provide an additional understanding of approaches and limitations alike.

The number of accidents in the Arctic is low. However, the consequences could be severe, owing to the presence of ice and other cold climate conditions, along with institutional factors such as remoteness and lack of preparedness infrastructure. The highest environmental risks concern the potential for fires on board the vessels and collisions with ice. The highest potential for loss of life are related to fires, grounding, and collisions with the ice for tourist ships and fishing vessels, along with the grounding and collision of cargo ships. The risks related to both grounding and collision with ice are significantly higher in the winter months. However, the number of ships is limited, and the vessels have a high ice class and/or are escorted by icebreakers. When it comes to the risk of violent actions, including terrorism, the probability is very low. The consequences can, however, be significant, owing to distances for medevac helicopters, the police, and special forces, with weather conditions such as fog hampering operations.

In Chapter 4, Dimitrios Dalaklis and Megan Drewniak reflect on search and rescue (SAR) capabilities in the Arctic. They raise questions about the risks and capacities of the Arctic countries to adequately respond to large-scale incidents. They claim that the existing emergency preparedness systems and response capabilities, including the number of ice-breaker emergency response vessels operating in the Arctic, are not sufficiently tested in large-scale accidents. The existence of adequate SAR centres and their performance are points of concern. Improved surveillance and communication systems are, among other things, in demand. Owing to limitations in SAR capabilities, it is important to look into the capacities and qualifications of seafarers in the region.

In Chapter 5, Bent Ove Jamtli takes this theme a step further and reflects on the increasing activity in the High North, both on land and at sea. He emphasizes the importance of an acceptable level of safety and preparedness for the sustainable development in the High North. He focuses closely on the challenges of the change in maritime traffic patterns and explores the icy and more remote regions and how this has implications for capacity improvements.

In Chapter 6, Frigg Jørgensen and Edda Falk provide interesting reflections in their commentary on how the cruise industry can supplement the governmental SAR and oil spill response system and provide additional emergency response capacities as vessels of opportunity. Expedition cruise ships sailing in Arctic waters could represent an excellent resource for preparedness and response, carrying equipment and a well-qualified crew. These vessels are often present in remote locations far from government SAR assets. The industry is co-operating closely with government emergency response agencies to increase mutual understanding and routines for survival and SAR and to disseminate competence.

Institutional Platforms for Risk Reduction and Emergency Response

International agreements and regulations can contribute to reduced risk, both through reducing the probability of unwanted incidents and increasing response capacities. As an example, the Polar Code established by the International Maritime Organization, with extra demands for the vessels, equipment, and competence, is a step in the right direction to increase robustness and to reduce the probability of an accident. Elgsaas and Offerdal (2018) focus on institutional arrangements and the potential for collaboration within the maritime preparedness systems of the Arctic to improve response capacities. They highlight the common interests of co-operation in the region, also improving emergency preparedness in the High North. A challenging task is navigating the complex systems of emergency preparedness within each country. There is a range of linguistic and cultural differences and divergent roles in international politics. Co-operation can be hampered by political conflicts, where less trust leads to less information exchange and hinders co-operation, mainly when military institutions and capabilities are involved. An important issue is the familiarization between countries with each other's emergency preparedness systems in the neighbouring countries of the High North.

Andreassen, Borch, and Schmied (2018) elaborate on the maritime emergency preparedness resources in the Arctic and capacity benefits from cross-border cooperation. In larger incidents, resource limitations represent a challenge for the leaders in charge of operations. A discussion on the joint exploitation of available physical and personnel capacities is important, including stationary facilities, specialized personnel, vessel capacities, airborne capacities, and management co-ordination capacity.

The need for additional resources that can be mobilized rapidly to deal with incidents outside the more densely populated mainland regions is difficult to estimate. There is a lack of systematic evaluations based on defined risk areas, clear response objectives, and capacity assessments. There are efforts towards achieving more frequent visits and the exchange and development of joint plans, systems, and procedures, among other things, through the Arctic Council's working groups and the Arctic Coast Guard Forum (ACGF). More studies of each organization's operational culture, shared operational systems, and IT tools can also provide more transparent co-ordination of resources.

Each nation's military preparedness system, including the navy and air force, represent significant capacity in the Arctic. Efforts to make military resources more available for civilian purposes could be a great opportunity, especially in the High North. Furthermore, the capacity of private co-operation, including in oil and gas, the cruise industry, local community resources, and other maritime activity, should be further assessed for preparedness operations, in order to give more insight on availability, mobilization time, and potential capacity. However, there is a risk of silo thinking and fragmented responsibility between institutions, companies, and organizations. Solutions for linking up organizations more closely and close co-operation on strategic, operational, and tactical levels are in demand.

In Chapter 7, Tor-Geir Myhrer claims that an open Arctic also makes the area more exposed to crime. This calls for the exchange of information and resources and close operational co-operation between the regional police services. A challenge is a lack of co-operation and significant differences in institutional structures when it comes to violent action and especially anti-terrorism response across borders. Myhrer shows that distances and legal challenges could hamper an efficient police response to crime. Outside a country's territorial waters, the opportunities for police intervention depend on the country's jurisdiction, for example defined by a maritime zone, then on what kind of crime is suspected, and lastly, what action is needed. So far, most maritime police work has focused on the illegal exploitation of resources, such as fishing, searching for oil and gas, and environmental crimes. One challenging area is organized crime and not least smuggling, including human-trafficking. Other challenges that are not likely but possible are the hijacking of ships and taking hostages, terrorist attacks on ships and oil/gas-producing installations, and piracy. Today, policing organization and procedures depend on the flag state. Myhrer claims that there is a need to amend existing multinational treaties in order to create a balance between the seriousness of the suspected crime and the local policing powers.

In Chapter 8, Rebecca Pincus illuminates how safety and security challenges provide the platform for building co-operative relations between the Arctic countries through their coast guards. The chapter focuses on the institutional aspects of co-operation between countries and describes the co-operative relations at different management levels. She highlights the institutional aspects of co-operation between several countries and how political tensions can influence co-operation across borders. The Arctic Coast Guard

Forum is claimed as a symbol of the value that the Arctic states place on good co-operation, which generates important macro-level benefits for all participants. The benefits include information-sharing on best practices and common challenges, sharing objectives on balancing activity and regulations, and achieving a good platform for ad hoc co-operation in the field. Special attention is brought to relationship-building, which helps to strengthen resilience against political conflicts that could threaten to undermine co-operation.

In Chapter 9 Captain Roberto H. Torres adds to the discussion with assertions that the ACGF is a bridge between diplomacy and operations. He discusses the way that the ACGF functions, and he claims that this forum is an ideal platform for the leaders and operators of the Arctic coast guards to share information, plan, carry out and learn from exercises. The ACGF enhances capabilities to collaborate effectively in the dynamic Arctic.

In Chapter 10, Jens Peter Holst-Andersen provides reflections on the difficult and complex environment that the Arctic represents. He emphasizes the need to meet and practice together. Co-operation between authorities is absolutely necessary to deal with situations like oil spills, large-scale incidents, and mass rescues in this vast and sparsely populated area. He draws on the examples of work of two international fora – The Arctic Council Working Group Emergency Prevention, Preparedness and Response and the ACGF as platforms for a higher level of interaction and a more competent and streamlined emergency response system.

Crisis and Emergency Management, Competence, and Training

Within each country and each emergency response field, there are differences in organizational models, responsibilities, and the main operational patterns. In large-scale operations, a broad range of agencies within SAR, oil spill response, firefighting, and violent action response might have to work together closely in a harsh maritime context. A central theme in this book is how to deal with variations in organizational structures, roles, responsibilities, command structures, and operational patterns.

Andreassen, Borch, and Ikonen (2019) reflect on how multi-sectoral and multinational co-operation can increase response capabilities in severe crises and in maritime areas where resources are scarce. The so-called host nation support system provides guidelines on how to access a broader range of resources from other countries. Challenges are present in providing knowledge of context, capabilities, and competence in response to large-scale incidents that seldom occur. Even though procedures are available for major disasters calling for mass evacuation, large-scale incidents such as 'black swans' events could demand significant resources in combinations not available within a single region or country. Combined operations with all the sectors mentioned above in the Arctic maritime domain will be complicated, owing to various command, control, and co-ordination systems between agencies and countries.

In Chapter 11, Svetlana Kuznetsova reflects on the co-ordination of oil emergency response, with examples from the Russian Arctic. Oil and gas activities in the Arctic create concerns about how to deal with possible accidents in the Arctic environment. Low temperature and harsh weather, a lack of transport infrastructure, and a shortage of oil booms and collection equipment adapted to freezing temperatures and high waves could hamper the oil spill response. Large-scale maritime oil spill incidents in the Arctic could result in an overload in the standard emergency response system. The chapter illustrates the potential strain on the co-ordinators who have to organize, monitor the operational environment, and share information that provides an overall situational awareness, despite information exchange limitations. Flexibility in the decision-making process is important at all management levels, including finding new resources and solutions, as well as adapting organizations and operating procedures to the prevailing environment.

In Chapter 12, Natalia Andreassen and Odd Jarl Borch take the organizational issue a step further and illuminate the importance of inter-organizational co-ordination performed by incident commanders in joint emergency response. The chapter shows that the range of managerial roles could change and have to be reflected on. The joint configuration of the integrated emergency organizations includes multiple actors across many jurisdictions, with diverging organizational design. In rapidly escalating situations, there is a need for immediate decisions and action that imply taking over authority from other levels and agencies. The authors show that the increased number of diverse elements to be allocated and controlled for demands managerial roles and mechanisms with more floating borders, flexible operating procedures, and tailor-made decision-making processes that are not hampered by rigid structures.

In Chapter 13, Jana Prochotska looks into the challenging task of counter-terrorism at sea and how to protect soft targets. The risk of terrorism calls for proper threat assessments to proactively build efficient human, material, partnership, and information capacities to address evolving security challenges. The author suggests defensive and proactive security measures to be considered. She emphasizes that broader regional co-operation on security issues is imperative, yet difficult to obtain without mutual trust between countries.

In several chapters, the need for competences and an adequate system for knowledge exchange has been highlighted. There is a need to understand the operational context of the Arctic, how the equipment and personnel are adapted, how to respond with limited resources available, and how to co-ordinate personnel belonging to a broad range of organizational systems, cultures, and professions. Improving the competence level through education, tailor-made courses, and exercises is an important issue in the emergency response system. In Chapter 14, Ensieh Roud and Johannes Schmied look into the benefits of collaboration on learning platforms, such as exercises. The chapter emphasizes the importance of training and especially full-scale exercises in preparing for emergency response in a maritime context.

In Chapter 15, Ole Kristian Bjerkemo emphasizes the importance of bilateral agreements on operating procedures and training. He illuminates the benefits of the oil spill agreement between Norway and Russia. A common command and control system for oil spill response out at sea has proved successful in exercises. He emphasizes, however, that combined operations such as an integrated sea and shoreline response will be much more complicated. The annual exercise stated in the bilateral agreement contribute to identifying improvements in joint contingency plans and operational guidelines at all decision levels.

In the commentary by Petr Gerasun in Chapter 16 the same issues are reflected upon within SAR operations. Each SAR operation in an Arctic environment is unique, and adaptations could be needed. It is very important to build and maintain the broad competences of the SAR mission co-ordinators at the maritime rescue co-ordination centres. The staff have to co-operate closely with the masters of the distressed vessel and need to have a proper understanding of the context and operational patterns of the vessels in the region.

In the final chapter of this book, Chapter 17, the editors reflect on the future knowledge needs for providing an efficient maritime response system for the Arctic. Many efforts have been made to provide platforms for knowledge exchange, common procedures, and joint exercises through institutional platforms such as the Arctic Council working groups and the ACGF. The International Maritime Organization is working on the qualifications and certification demands for the vessel's crew and maritime SAR routines. Bilateral agreements between countries such as Norway and Russia facilitate annual exercises both within SAR and oil spill response, exposing knowledge gaps. An increased focus on competencies related to organizational structures, processes, culture, operational procedures, management, and exercises are needed, especially related to mass evacuation and operations in icy waters. Furthermore, a focus should be placed on the joint response to violent actions and counter-terrorism efforts in the Arctic region. Laws, regulations, and institutions, as well as political relations between the whole range of stakeholders in the Arctic, need more scrutiny from several research disciplines.

References

Andreassen, N., Borch, O.J., and Ikonen, E.S. (eds.) (2019). Organizing emergency response in the European Arctic: a comparative study of Norway, Russia, Iceland and Greenland, MARPART Project Report 5. Bodø: Nord Universitet 2019, R&D report N46. http://hdl.handle.net/11250/2611539.

Andreassen, N., Borch, O.J., and Schmied, J. (eds.) (2018a). Maritime emergency preparedness resources in the Arctic: capacity challenges and the benefits of cross-border cooperation between Norway, Russia, Iceland and Greenland. MARPART Project Report 4, R&D report 33, Nord university. http://hdl.handle.net/11250/2569868.

Borch, O.J., Andreassen, N., Marchenko, N., Ingimundarson, V., Gunnarsdóttir, H., Iudin, I., Petrov, S., Jakobsen, U., and Dali, B. (2016a). Maritime activity in the High North - current and estimated level up to 2025. MARPART Project report 1. Bodø: Nord Universitet Utredning (7) 130s. http://hdl.handle.net/11250/2413456.

Borch, O.J., Andreassen, N., Marchenko, N., Ingimundarson, V., Gunnarsdóttir, H., Jakobsen, U., Kern, B., Iudin, I., Petrov, S., Markov, S., Kuznetsova, S. (2016b). Maritime Activity and Risk Patterns in The High North. MARPART Project Report 2. Bodø: Nord Universitet 2016 (ISBN 978-982-7456-7757-3) 124 s. Nord universitet / R&D report (4). http://hdl.handle.net/11250/2432922.

Campbell, D.J. (1988). Task complexity: A review and analysis. Academy of Management Review, 13: 40–52.

Czarniawska, B. (2004). On time, space, and action nets. *Organization*, 11: 773–791.

Elgsaas, I.M. and Offerdal, K. (eds.) (2018). Maritime preparedness systems in the Arctic – institutional arrangements and potential for collaboration. MARPART Project Report 3, FoU-rapport 27, Nord University. http://hdl.handle.net/11250/2501164.

Gephart, R.P., Miller, C.C, and Helgesson, K.S. (2019). *The Routledge Companion to Risk, Crisis and Emergency Management*. New York/Oxford:Routledge.

Hærem, T., Pentland, B.T., and Miller, K.D. (2015). Task complexity: Extending a core concept. *Academy of Management Review*, Vol. 40, No. 3, 446–490.

Johannessen, S.O. (2018). *Strategies, Leadership and Complexity in Crisis and Emergency Operations, Routledge Advances in Management and Business Studies*. New York: Routledge.

Kahneman, D. and Klein, G. (2009). Conditions for intuitive expertise: A failure to disagree. *American Psychologist*, 64(6), 515–526.

Steigenberger, N., Lubcke, T., Fiala, H.M., and Riebschlager, A. (2017). *Decision Modes in Complex Task Environments*.

Tversky, A. and Kahneman, D. (1974). *Judgement under Uncertainty: Heuristics and Biases. Science*, Vol. 185, no. 4157, 1124–1131.

Weick, K.E. and Sutcliffe, K.M. (2011). *Managing the Unexpected: Resilient Performance in an Age of Uncertainty*. Chichester: John Wiley & Sons.

Wolbers, J., Boersma, K., and Groenewegen, P. (2017). Introducing a Fragmentation Perspective on Coordination in Crisis Management. *Organization Studies*, 1–26.

2 The Opening of the Arctic

How Newcomers Gain Access to a Complex Environment

Torbjørn Pedersen

Introduction

The Arctic region is attracting international attention as ever before. A number of nations, Arctic and non-Arctic alike, have in recent years developed comprehensive Arctic policies and strategies (Heininen et al,. 2020) and take aim to increase their presence and involvement in the region, for security and/or economic reasons (ibid.). One widely shared conception is that climate change and receding sea ice are making natural resources and new trans-Arctic sea lanes more accessible (Borgerson, 2008; Heninen et al., 2020). Amid increased attention, newcomers are expected to increase their presence in a region known to have been one of the most inhospitable and exclusive operation areas on Earth (Norwegian Intelligence Service, 2019).

During the Cold War, the USA and the Soviet Union competed to acquire the better knowledge of the Arctic region (Central Intelligence Agency, 1984; Wells, 2017) − a sensitive operational area wedged between the two global super powers. '[T]hat navy possessing a superior knowledge of the environment and knowing how to take tactical advantage will be the victor', Admiral of the Soviet Fleet, Sergey Gorshkov, once stated (Åtland, 2007). Unique knowledge of the elements − such as sediments of the seabed, bathymetry, ocean physics, meteorology, or the ionosphere − could translate directly into a pronounced advantage in any given military encounter between the Cold War rivals (Pedersen, 2019; Winters, 1998).

The Arctic was exclusive, in the sense that few other nations had the necessary knowledge, not to say platforms, to operate in this complex and hostile environment. Today, the region is becoming more inclusive, this chapter argues. Newcomers − notably China − are rapidly catching up with the traditional powers, owing in part to their own extensive polar research, but, even more so, because of massive research collaboration internationally in various fields and the extensive sharing of Arctic research and on-site data.

This chapter describes the proliferation of knowledge of the Arctic natural elements and discusses its consequences for both civilian and military stakeholders. The conclusion is twofold: As the traditional knowledge gap closes, new actors, including Chinese civilian as well as military operators, will no longer be at a distinct operational disadvantage in this hostile environment. It makes civilian navigation safer, but it also allows for the presence of naval vessels from non-Arctic states and the subsequent tension between old and new stakeholders.

This chapter focuses on access to the maritime Arctic. It also pays special attention to newcomer China, given its impact on the world order as a rising great power (e.g. Mearsheimer, 2014; Allison, 2017), its aspirations to develop a trans-Arctic Polar Silk Road (e.g. China State Council, 2018; Brady, 2017), and its maritime relevance as one of the world's top flag states, with more than 6,000 cargo vessels above 500 gross tonnage registered in China and Hong Kong combined, as at 2018 (Lloyd's List Intelligence, 2018).

Theory

Knowledge about the elements and the ability of an operator to predict their future state can be expressed in terms of situation awareness. Mica Endsley defines situation awareness as 'the perception of the elements in the environment within a volume of time and space, the comprehension of their meaning, and the projection of their status in the near future' (Endsley, 1995: 26). The definition implies three levels of situation awareness, moving up to higher levels as the operator increases his or her ability to predict the state of the surrounding environment and hence take more qualified decisions. While Endsley takes a predominantly user-centered approach to situation awareness, relating the three levels to cognitive processes and the individual's ability to make decisions in a highly dynamic environment (Endsely and Jones, 2012), the three-level terminology can also be related to the sum of systems, data assimilation schemes and underlying data necessary for operators to navigate safely and effectively in a dynamic maritime environment (Pedersen, 2019).

Maritime operations depend on a number of systems and prediction models, including weather forecasting, ocean circulation models, wave, surf and tide models, ice predictions, and atmospheric models (National Research Council, 2003).

High-resolution maps and accurate systems, which provide continuously updated forecasts of the operators' dynamic environment, are essential to safe navigation in the Arctic maritime domain. Here, in the extreme north, the environment is particularly vast, harsh, and isolated (Christensen, 2010). Reliable systems, fed with massive amounts of historic, on-site, and remotely sensed data, are therefore a crucial precondition for any Arctic operator's situation awareness. The complexity of the best systems is immense. The US Navy Environment Prediction System, a state-of-the-art system, combines 4-Dimensional Variational Data Assimilation (4D-VAR) with numerous ocean circulation models, wave, surf and tide models, ice models, and atmospheric models (Burnett et al. 2014). Sensors, both on-site and remote, continuously feed the data assimilation schemes, allowing the system to provide real-time, high-resolution, and three-dimensional models and predictions of the operational environment (ibid.).

In military operations, as already noted, a superior knowledge of the environment could translate into a tactical advantage (Pedersen, 2019; Winters, 1998). This is perhaps most evident in undersea warfare, where stealth is key to survival (Polmar and Whitman, 2016). Undersea warfare not only relies on underwater terrain models, complex ocean circulation models, and acoustic propagation predictions. Antisubmarine warfare also combines a range of non-acoustic detection systems, as a submarine also could be detected by its electric, magnetic, seismic, and pressure influences (SAES, 2017).

This adaption of the three-level situation awareness terminology, stressing the need for systems, assimilation schemes, and relevant data, should not be confused with the concepts of team, or shared, situation awareness (e.g. Salmon, 2008). As systems, schemes, and data from a given region are shared worldwide, operators are likely to achieve a more aligned – and higher degree of – situation awareness, as the knowledge gap between nations and operators narrows.

Methodology

As noted, the US National Research Council (2003) has named a number of elements that strongly affect marine operations in the Arctic. A previous study (Pedersen, 2019) has detailed how the various elements impact navigation safety. The most relevant maritime elements fall into three interacting spheres: the atmosphere, the ocean, and the seabed and subsoils.

This chapter outlines some of the most significant international research efforts and research collaborations related to each sphere, in addition to the interaction between them. The efforts and collaborations have been identified through a mix of methods, including informal interviews, archival and database searches, and document analyses, to bring additional reliability and validity to the findings (Yin, 2003).

International polar research includes thousands of researchers from dozens of states, and research collaboration can be elusive, as it is often small-scale, seminar-like, and/or arranged in informal networks across regions (Pedersen, 2019). Without being unappreciative of these myriad interactions, this study primarily aims to identify research efforts and international collaborations that yield significant knowledge proliferation across borders.

This chapter does not claim to provide an exhaustive overview of all relevant research and collaborations, but, rather points to some significant structures and processes through which knowledge proliferates.

Results

The secrets of the Arctic are uncloaked as international researchers flock to the top of the world to learn about its elements amid global warming, which is most pronounced here (Norwegian Polar Institute, 2018. The number of nationalities involved in polar research has surged in recent years. Projects related to the International Polar Year 2007–2009 involved researchers from more than 60 nations (International Polar Year, 2010). Currently, the International Arctic Science Committee, a non-governmental organization facilitating Arctic research collaboration, counts 23 members, which is three-fold since 1990. Among its members are China, India, Japan, and South Korea (Barr, 2015).

Some of the most capable ice-reinforced research vessels operating in the Arctic today fly the flags of non-Arctic nations, including China. On the MV *Xue Long*, Chinese researchers have mapped and developed comprehensive sailing manuals for both the Northwest and Northeast Passages (Pedersen 2019). China has also experimented with large stationary buoys, deployed to monitor high-altitude, real-time meteorological and oceanographic parameters, including conductivity, temperature, and density (Pedersen, 2019). In 2020 China's capacity to operate in the polar regions is greatly enhanced, as MV *Xue Long 2* – a Polar Class 3 icebreaker – has entered service.

Weather forecasting in the Arctic has long been challenging. On-site observations have been few, and traditional meteorological models have not been designed with Arctic conditions in mind (Pedersen, 2019). However, the forecasting challenges are now diminishing as nations increasingly share their observations and develop common weather models (ibid.). Specialized international collaborations include the World Meteorological Organization, which has initiated a Polar Prediction Project (PPP) aimed at improving weather prediction services in the region through international collaboration. Here, China is closely involved and sits in the steering committee of Year of the Polar Prediction, a flagship activity of PPP aimed to enable a significant improvement in environmental prediction capabilities for the polar regions and beyond, by coordinating a period of intensive observing, modelling, verification, user-engagement, and educational activities (Polar Prediction, 2018).

China is also participating in space weather research and data-sharing in the Arctic, close to the magnetic cusp, where the effects of space weather are particularly pronounced (Rose and Ziauddin, 1962). Space weather affects a range of systems. According to H.C. Koons et al., 'known impacts include service outages, mission degradation and mission failure, data loss, sensor degradation, subsystem failure, launch delays, redesign and retest, anomaly analyses, and the ultimate cost for each of the preceding' (Koons et al., 1999: 1). Newcomer China has invested in the China-Iceland Joint Arctic Science Observatory at Karholl, Iceland and become a full partner of the European Incoherent Scatter radar collaboration outside Longyearbyen, in the Norwegian islands of Svalbard.

Oceanographic systems, models, and data are also widely shared across nations. A notable research co-operation for European oceanographers is the Arctic Region Ocean Observing System (ArcticROOS), a regional collaboration of the European Global Ocean Observation System (EuroGOOS), set to include members from outside Europe (Pedersen, 2019). On-site observations, from various platforms and actors, are already made openly available to international researchers. Data from the Hausgarten deep-water observatory in the Fram Strait, for instance, are shared by Germany's Alfred Wegener Institute online. Likewise, the FerryBox system, operated by members of EuroGOOS, is installed on board vessels sailing regularly along the Norwegian coast and between Norway's mainland and Svalbard, and is uploading a stream of real-time oceanographic data to researchers worldwide. At the same time, oceanographic systems and models are becoming more available commercially off-the-shelf and sophisticated as data-processing power increases and the end-user market grows.

High-resolution models and maps of the Arctic seafloor, once guarded as national secrets, are also shared by state and non-state actors alike through various programmes. Norway's multi-year Mareano research programme, for instance, has provided and shared a treasure trove of details about the bathymetry, sediments, and biology of the Norwegian continental shelf. Non-state actors are also accumulating details about the seafloor through data-sharing. The Olex echo-sounder allows its users to accumulate an increasingly detailed and comprehensive 3D map of the seafloor.

In total, international collaboration in polar research extends from classroom-sized seminars to large and complex expeditions, such as the 2019–2020 Multidisciplinary drifting Observatory for the Study of the Arctic Climate (MOSAiC), an expedition into the central Arctic Ocean led by the Alfred Wegener Institute involving researchers from 16 different nations across Europe, Asia, and North America. Co-operation involves a range of state and non-state actors. Data from various monitoring programmes, expeditions, and national databases are increasingly made available by state as well as non-state actors through various databases, such as Copernicus, EMODnet, and Pangaea.

Discussion and Conclusion

Shipping in the Arctic is increasing, for various reasons. For one, Russia has named the Arctic region its 'primary resource base' for the 21st century (Åtland, 2011), precipitating Russian investments in natural resources along the northern coast. A Liquid Natural Gas plant in the Yamal Peninsula is already generating a spike in ship movements in the Kara and Barents Sea (source). Also, offshore petroleum production has been introduced to the region, with the opening of the 'Snøhvit', 'Goliat', 'Prirazlomnoye', and 'Johan Castberg' fields in recent years, and Arctic cruise tourism is gaining momentum. Fishing remains a significant activity, as the region boasts some of the largest and most robust fish stocks in the world (Hønneland and Jørgensen, 2018). Some shipowners are also probing the Northeast Passage as an alternative route between Asia and Europe amid the loss of sea-ice in the Arctic Ocean (Lasserre et al., 2016).

'[T]he need for adequate and accurate environmental data on small scale is paramount for minimizing uncertainty and reducing risk', according to the National Research Council in a report on environmental information for naval warfare (National Research Council, 2003). The same could be said about civilian shipping. Shipping in extreme latitudes calls for the proliferation of high-resolution and real-time knowledge of the Arctic elements and high-level situation awareness of all maritime operators. With the higher levels of situation awareness, which, according to Endsley's terminology, implies the ability to predict the future state of the elements (Endsley, 1995), operators are better equipped at safeguarding human life and property at sea, as well as a fragile ecosystem.

In the context of Arctic navigation, access to data – historic, on-site, and remote – is crucial to all systems designed to provide accurate readings and predictions of the environment. With more nations getting involved with polar research, and with massive research collaboration and data-sharing across national borders, this data is becoming more widely distributed and accessible to all who seek it. High-resolution prediction systems, as well as meteorological and oceanographic models, have become commercial off-the-shelf. What used to be highly classified knowledge about the Arctic elements, which only a handful of military stakeholders benefited from, is becoming widely accessible, both to civilian and military operators. Situation awareness is likely to be more evenly shared across nationalities and operators as knowledge of the elements proliferates.

Furthermore, knowledge and activity are interlinked and interdependent. More knowledge facilitates the situation awareness necessary to operate safely in the region. And with more activity, knowledge is accumulated. Knowledge enables activity, and activity generates knowledge.

The increased ability of newcomers to operate in the Arctic also extends to their armed forces. According to the US Department of Defense (2018) and the Norwegian Intelligence Service (Lunde, 2019), China will soon be capable to operate naval vessels in a region that used to be exclusive to the chief Cold War rivals. With higher levels of situation awareness, coupled with more capable hardware, China will soon pursue its interests in the Arctic region by military means as well. These interests include freedom of navigation, but also access to natural resources in the region. Thus, as polar research, international research co-operation, and data-sharing raise situation awareness across nationalities and make Arctic shipping safer, they also facilitate the military presence of newcomers to the region, potentially adding tension to the region.

References

Allison, Graham. (2017). *Destined for War: Can America and China escape Thucydides's Trap?* (Boston, MA: Houghton Mifflin Harcourt)

Borgerson, Scott. (2008). Arctic Meltdown: The Economic and Security Implications of Global Warming. *Foreign Affairs.*

Burnett, William et al. (2014). Overview of Operational Ocean Forecasting in the US Navy Past, Present, and Future. *Oceanography*, 7, 24–31.

Central Intelligence Agency. (1984). The Soviet Oceanographic Research Program. A Technical Intelligence Report SW 84–10007 (declassified 5 March 2010).

China State Council. (2018). "China's Arctic Policy", 26 January.

Christensen, Kyle. (2010). The Arctic: The Physical Environment (Defence R&D Canada) DRDC-CORA TM 2010–193.

Endsley, Mica. (1995). Toward a Theory of Situation Awareness in Dynamic Systems. *Human Factors Journal*, 37: 26.

Endsley, Mica and Jones, Debra G. (2012). *Designing for Situation Awareness: An Approach to User-Centered Design*. (Boca Raton, LA; London and New York: CRC Press).

Heninen, Lassi et al. (2020). Arctic Policies and Strategies – Analysis, Synthesis and Trends. International Institute for Applied Systems Analysis.

Hovem, Jens. (2012). *Marine Acoustics: The Physical Sound in Underwater Environments*. (Los Altos Hills, CA: Peninsula Publishing).

Hønneland, Geir and Jørgensen, Anne-Kristin. *Implementing International Environmental Agreements in Russia*. (Manchester: Manchester University Press).

International Polar Year. (2010). About IPY. https://ipy.arcticportal.org/about-ip.

Koons, H.C. et al. (1999). The Impact of the Space Environment on Space Systems. *Aerospace Report*, 99(1670)-1.

Lassere, Frederic et al. (2016). Polar Seaways? Maritime Transport in the Arctic: An Analysis of Shipowners' Attentions. *Journal of Transport Geography*, 57, Lloyd's List Intelligence. https://lloydslist.maritimeintelligence.informa.com/LL1125024/Top-10-flag-states-2018.

Lunde , M. (2019). Presentasjon Fokus. Oslo Militære Samfund, 11 February.

Mearsheimer, J. (2014). *The Tragedy of Great Power Politics* (revised ed.). (New York: W.W. Norton & Co.).

National Research Council. (2003). *Environmental Information for Naval Warfare* (Washington, DC: National Academies Press).

Norwegian Intelligence Service. (2019). Fokus 2019.

Norwegian Polar Institute. (2018). Klimaendringer i Arktis. www.npolar.no/tema/klimaendringer-arktis/.

Pedersen, Torbjørn. (2019). Polar Research and the Secrets of the Arctic. *Arctic Review on Law and Politics*, 10.

Polmar, Norman and Whitman, Edward. (2016). *Hunters and Killers, Vol. II: Anti-Submarine Warfare from 1943*. (Annapolis, MD: Naval Institute Press).

Rose, D.C. and Ziauddin, Syed. (1962). The Polar Cap Absorption Effect. *Space Science Review* 1, 115–134.

SAES. (s.a.). SEAPROF – Undersea Acoustic Performance Prediction System. Accessed 25 October 2018. https://electronica-submarina.es/wp-content/uploads/2018/08/SAES_SEAPROF_Acoustic_Model_english.pdf.

Salmon, Paul et al. (2008). What really is going on? Review of situation awareness models for individuals and teams. *Theoretical Issues in Ergonomics Science*, 9, 297–323.

Sun Tzu. (2014 edition). *The Art of War*. (London: Penguin Books), 81.

Wells, Anthony. (2017). *A Tale of Two Navies: Geopolitics, Technology, and Strategy in the United States Navy and the Royal Navy, 1960–2015*. (Annapolis, MD: Naval Institute Press).

Wells, Anthony. (2017). *A Tale of Two Navies: Geopolitics, Technology, and Strategy in the United States Navy and the Royal Navy, 1960–2015.* (Annapolis, MD: Naval Institute Press, 2017).

Winters, Harold. (1998). *Battling the Elements: Weather and Terrain in the Conduct of War.* (Baltimore, MD: Johns Hopkins University Press).

World Meteorological Organization. (2017). Year of Polar Prediction – From Research to Improved Environmental Safety. Press release, 15 May.

Yin, R.K. (2003). *Case Study Research. Design and Method.* (Thousand Oaks, CA: Sage Publications).Young, O. (1992). *Arctic Politics: Conflict and Cooperation in the Circumpolar North.* (Lebanon, NH: University Press of New England).

Åtland, Kristian. (2007). The Introduction, Adoption and Implementation of Russia's 'Northern Strategic Bastion' Concept, 1992–1999. *The Journal of Slavic Military Studies*, 20, 499–528.

Åtland, Kristian. (2011). Russia's Armed Forces and the Arctic: All Quiet on the Northern Front? *Contemporary Security Policy*, 32, 267–285.

3 Maritime Activity and Risk in the Arctic

Nataliya Marchenko

Introduction

For many centuries, the Arctic was regarded as cold, dark, remote region suitable only for hunter and adventurers. In the 21st century, the Arctic is the place with huge and attractive mineral recourses and opportunities for new navigational routes and tourism. Sea-ice shrinking and a lack of resources in the regions with a moderate climate led to the substantial growth of activity. Increased activity could lead to more accidents and requires greater preparedness for emergencies and a change in the regulation of marine navigation. It requires an adequate estimation of risk. Among the various methods of risk assessment, there are quite a few that apply to Arctic navigation, owing to a lack of statistics for rare events such as marine accidents in these remote waters. Furthermore, there is a lack of data and visualizations of those risk assessments for effective management for risk mitigation. As a result, qualitative methods are needed for the Arctic.

How can we assess the risk and make our assessments practical and applicable in the current conditions? How can we utilize the knowledge on handling marine emergencies in the Arctic and use this experience to be useful for the prevention of future losses?

The Main Maritime Activities

To assess risk, there is a need to investigate the maritime traffic patterns of the Arctic. The main maritime activities in the Arctic are presented in Figure 3.1, where fishing is denoted in grey areas, hydrocarbon exploration and production are in dark grey, and cities are grey circles; transport routes are shown as lines (existing routes in grey lines and planned routes in black lines). The density of maritime traffic is presented in Figure 3.2.

There are broadly five fields of activity in Arctic waters, connected to a specific type of vessels. In all these fields, the level of activity is growing, and the movement to the North is remarkable (Borch, Andreassen, et al., 2016a; Borch, Andreassen, et al., 2016b).

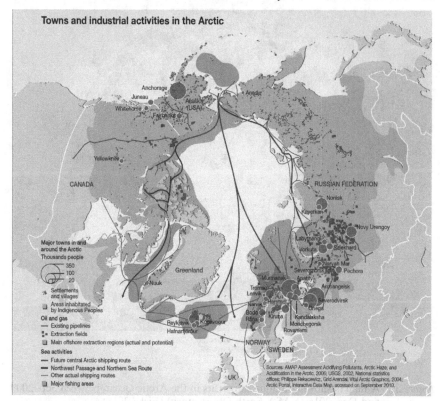

Figure 3.1 Towns and activities in the Arctic. GRID-Arendal graphic (Pravetoni 2010).

1) **Transport/cargo.** Arctic shipping has increased noticeably in recent years in all regions. Marine traffic in the Canadian Arctic almost tripled in 1990–2015; the annual distance travelled by all vessels grew from around 350,000 km to over 900,000 km, with the majority of growth occurring over the past decade (Dawson, Pizzolato, et al. 2018). Russian Arctic shipping more than quadrupled in five years, and huge investments are expected to increase traffic on the Northern Sea Route even more (Khon, Mokhov, et al., 2010), (CHNL IO, 2019). Transport vessels are getting bigger and often carry dangerous goods. They usually have a large amount of fuel, not to mention oil tankers. Such vessels are dangerous for the environment in the case of an accident, followed by an oil spill.

2) **Tourism.** The Arctic is becoming more and more attractive for large cruise vessels and for smaller adventure ships. Port calls in Longyearbyen have almost tripled during the past ten years. Cruises along the Northeast and Northwest Passages, to the North Pole on Russian nuclear icebreakers are performed on a regular basis.

Cruise vessels are the primary safety concern when it comes to risk to life because of the large number of passengers and huge potential consequences.

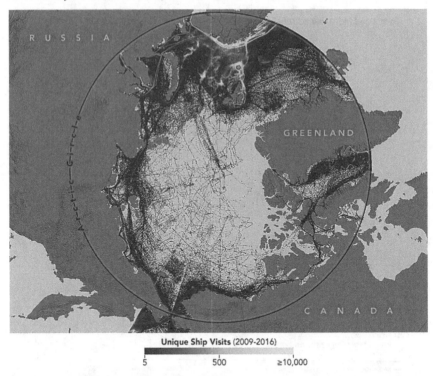

Unique Ship Visits (2009-2016)

5   500   ≥10,000

Figure 3.2 A map is showing unique ship visits in the Arctic Ocean between Sep. 2009 and Dec. 2016 (NASA Earth Observatory) (Fiske 2018).

In general, these tourists do not consider the consequences of unforeseen circumstances. Even travel agencies and insurance companies do not adequately assess and calculate possible risks.

There are not enough resources in the tiny Arctic communities to assist and accommodate people in the case of a cruise ship accident with 3,000–4,000 tourists on board. In the harbour of the largest and northernmost town of Longyearbyen, Svalbard, Norway, with a population of approximately 2,300 people, huge tourist ships come almost every week in the summer season. These ships go further north, moving away from settlements for tens and hundreds of kilometres.

3) **Fishing.** Higher winter temperatures have impacted fishing, causing expansion northwards and eastwards where there was previously little or no trawling. With reduced sea-ice extent, vessels fishing with bottom trawlers can travel further north than before, entering areas that have previously were not of interest to fishermen. Fishing vessels are mostly small in size, but they are well represented in accident statistics and risk assessments because of their significant numbers. Relatively few fishing vessels dock in the remote ports like Longyearbyen, so they are not noticeable in port statistics. However, fishing vessels are evident on automatic identification system (AIS) maps,

producing a thick web of tracking lines that is present in all seasons and even in the long, dark winter and the spring with sea-ice, streaming far north.

4) **Mineral resources exploration and exploitation.** Currently, it is mostly oil and gas resources that are actively being developed in the Arctic, containing more than one-fifth of the world's reserves. However, other resources are becoming more important. New hydrocarbons exploration fields are located further and further north – near Bear Island in the Barents Sea (74°N) and in the Kara and Laptev Sea. Drilling and supply vessels are usually medium-sized and are well prepared for unexpected situations, as they can help in search and rescue (SAR) operations. Offshore oil constructions are potentially dangerous for the fragile Arctic environment.

5) **Research and monitoring (including military).** These types of vessels are pioneers in remote and unexplored waters. They are medium-sized and have advanced equipment. The main concern for this type of vessel is the considerable distance from 'civilisation' that they can reach, making it a challenging task to send help in the case of an emergency.

The first three types of vessels connected to the aforementioned activities will be under consideration in the risk assessment in this chapter, as they form the main concern in terms of health and safety.

All these activities are visible on the map compiled using AIS system (Figure 3.2). The maritime activities cover the Arctic in a rather thick net of lines. It is evident that in the Atlantic sector of the Arctic all shipping is becoming more active, going further north, driven by warm North Atlantic currents, affecting sea-ice, which is the main limiting factor for navigation in the Polar region. According to Fiske (2018), the Norwegian marine sector accounts for some 80% of all shipping activities.

In this study, we concentrate on the Atlantic sector, as a prototype of the future development of the entire Arctic in the case of global warming.

Risk Assessment Theory and its Applicability in the Maritime Arctic

The traditional definition of risk is emphasising the estimated amount of harm that can be expected to occur during a given period, owing to a specific event. For example, the UN Office for Disaster Risk Reduction determines disaster risk as 'The potential loss of life, injury, or destroyed or damaged assets which could occur to a system, society or a community in a specific period, determined probabilistically as a function of hazard, exposure, vulnerability and capacity' (UN Office for Disaster Risk Reduction, 2009–17). The International Organization for Standardization (ISO) in the document *ISO 31000, Risk management – Guidelines* defines risk as the 'effect of uncertainty on objectives' (International Standardization Organization, 2018). This document implies the indeterminacy of event possibility, ambiguity, or a lack of information. So, risk is the product of the probability that an accident happens multiplied by the adverse effects on health, environment, and values that an accident could cause.

A typical risk matrix has rows representing the increasing likelihood of an accident to occur and columns representing increasing the severity of consequences of a hazard (Trbojevic, 2000). On a standard risk matrix, red cells indicate high risk, yellow ones moderate, and green ones low (see risk matrix for Svalbard as an example, Table 3.2). After identifying risk, measuring it, and estimating the consequences, a traditional risk management process encourages a response, which involves, among other things, a risk mitigation strategy (Crouhy, Galai et al., 2006).

The risk matrix approach has its limitations, owing to the difficulties of quantifying the components of risk and their possible correlation and the limited ability to reproduce risk ratings accurately (Cox Jr, 2008). In spite of these limitations, the risk matrix approach is widely used for many practical estimates. For example, risk matrices were utilized during initial discussions on preparedness improvement, because they provide a coarse-grained picture of risk levels as a basis for further assessments. They also serve as a platform for a discussion on priority needs both regarding precautions and safety efforts and the allocation of preparedness resources. The Polar Code established by the International Maritime Organization for icy waters also started with risk matrices. The Polar Code has become a significant step towards reducing the probability of accidents through more robust vessels and improved training, as well as a step towards reducing the consequences of an accident for better SAR preparedness on board vessels (International Maritime Organization, 2014).

However, the assessment of risk in the Arctic sea regions is a challenging task because the conditions are changing, and there is a lack of incident statistics for calculating probabilities. That is why we can find, in practice, two various types of risk assessment: 1) local and detailed; and 2) regional and unified. The first one is performed for a particular area, often map-based and gives a risk assessment of a particular task in a defined time. Dedicated authorities perform such assessments using the full range of current data.

The best example of such assessments in Arctic waters is the Canadian Arctic Shipping Risk Assessment System (CASRAS) (Kubat, Charlebois, et al., 2017). CASRAS estimates route navigability in the Canadian Arctic, using the Polar Operational Limit Assessment Risk Indexing System (POLARIS) for a vessel, according to the observed or predicted ice conditions. For example, in Figure 3.3, there is the assessment of 14 March 2016, for the route to one of the North Gulf on Baffin Island (black line) on the base of the ice chart. Yellow and orange zones require a reduction in speed and escort vessels. This approach allows for the use of quantitative analysis, using available numerical data (weather, ice, number of vessels in the area, number of animals, protected species, etc.). The same approach was used in the assessment of environmental risk connected to potential oil pollution from ship traffic in seas around Svalbard and Jan Mayen, risk assessment regarding

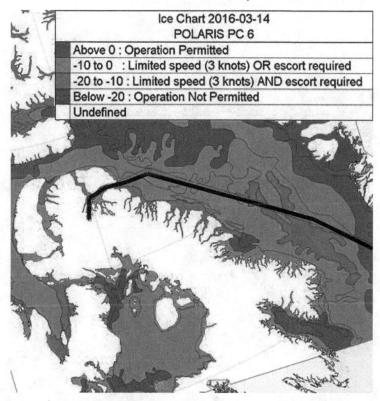

Ice Chart 2016-03-14 POLARIS PC 6	
	Above 0 : Operation Permitted
	-10 to 0 : Limited speed (3 knots) OR escort required
	-20 to -10 : Limited speed (3 knots) AND escort required
	Below -20 : Operation Not Permitted
	Undefined

Figure 3.3 CASRAS (Canadian Arctic Shipping Risk Assessment System) example (Kubat, Charlebois et al. 2017).

a piloting service on Svalbard, and the risk of oil pollution for protected species and areas (DNV GL, 2014, DNV GL, 2014).

Some of the assessments are presented in the form of GIS. For example, 'The Arctic risk map' web application, created by DNV GL, an international accredited registrar and classification society (DNV GL, 2019); the HELCOM portal (HELCOM, 2019) created by the Baltic Marine Environment Protection Commission – Helsinki Commission, depicting the risk of oil spills from vessels grounding, modelled for 2008/2009 (Figure 3.4).

Recently, several investigations on a quantitative approach in risk assessment were implemented. For example, the paper 'Risk analysis of offshore transportation accident in Arctic waters using Markov Chain Monte Carlo framework' (Abbassi, Khan, et al., 2017) is an attempt to quantify the probability of a ship accident, building detailed fault trees for three possible scenarios, considering factors related to cold and harsh conditions and their effects on grounding, foundering, and collision.

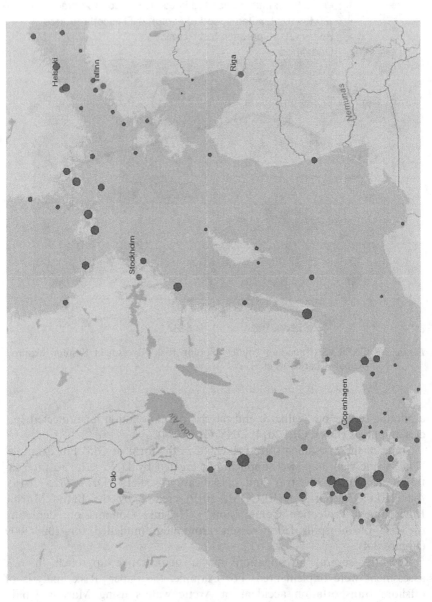

Figure 3.4 HELCOM Risk Assessment. Risk of an oil spill from grounding. The area of the bubbles corresponds to the risk of spill of oil and hazardous substances. The unit of the risk is average tonnes per year. Made in (HELCOM 2019) using the layer (Lentz 2013).

The methodology is implemented in a Markov Chain Monte Carlo framework to assess the uncertainties that arise from historical data and expert judgements involved in the risk analysis. The study 'Marine transportation risk assessment using Bayesian Network: Application to Arctic waters' (Baksh, Abbassi et al., 2018) proposes the risk model, considering different operational and environmental factors (ice as the dominant factor) and illustrates it through a case study of an oil-tanker navigating the Northern Sea Route in the Russian Arctic. By running uncertainty and sensitivity analyses of the model, a significant change in the likelihood of the occurrence of accidental events is identified. The investigation 'A quantitative approach for risk assessment of a ship stuck in ice in Arctic waters' (Fu, Zhang, et al., 2018) presents a frank copula-based fuzzy event tree analysis approach to assess the risks of major ship accidents in Arctic waters, considering uncertainty. The quantitative approach includes three steps: accident scenario modelling by an event tree model, probability and dependence analysis of the associated intermediate events, risk assessment concerning the following outcome events. A major ship accident in Arctic waters – ships stuck in ice, is chosen as a case study to interpret the modelling process of the approach proposed.

However, extensive, reliable numerical data to perform quantitative analysis is available only for specific situations, on a local scale for small areas. All existing quantitative models can be run for past events, where all parameters are known. If we need to estimate the region in general for various types of events, we have to use the second type of assessment, based on qualitative and expert estimation. That is why we introduced and developed the method of risk assessment for large regions and typical emergencies, using analysis of previous accidents and trends in shipping development. The method will be described in the next section.

The Method of General Risk Analysis for Large Regions

As shown above, quantitative risk assessments can be performed for particular (not so large) areas and a particular event in the limited period, because the conditions (i.e., ice, weather) and factors (ship traffic, technology, and so on) are constantly changing. For the development of preparedness systems in the Arctic, and management of emergency response resources, the general risk assessment for large regions should be performed.

A risk assessment made for several sea areas in the High North region (Norway and Russia west of Novaya Zemlya) provides an illustration of differences in risk between regions and types of activity (Marchenko, Borch, et al., 2015). (Marchenko, Borch, et al. 2016) reflected on available SAR resources, their compliance with risk assessment, and identified capacity gaps. The analysis was expanded to the west (Iceland and Greenland) and includes the analysis of five large regions of Atlantic Arctic (Figure 3.5)

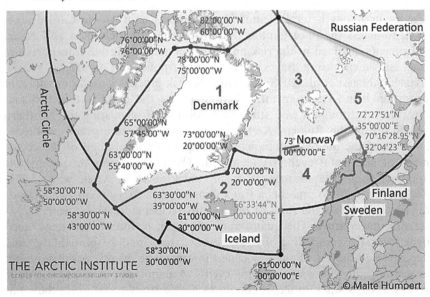

Figure 3.5 Regions considered in risk assessment: 1-Greenland, 2-Iceland, 3-Svalbard, 4-Coastal Norway, 5-Russian sector of the Barents Sea. The base map is the Arctic Search and Rescue Agreement Map (Arctic Council, 2011).

(Marchenko, Andreassen, et al., 2018) and elaborates on the range of challenges related to remoteness, the risk of ice and icing, and limited government resources.

The risk assessment procedure for large regions (e.g., Svalbard Area – number 3 in Figure 3.5), averaged over the significant period when accidents happen (e.g., on average during the year over the past ten years; the forecast scenario for years ahead can also be used) is described below.

In the first step of risk assessment, we define the most relevant types of accidents as the combination of the type of ship, the operation in the area (first letter in abbreviation), and the specific type of emergency (second letter in the abbreviation), as shown in Table 3.1. The main ship types in for Arctic waters (tourist, cargo, and fishing) are described shortly. The main emergencies are grounding (meaning that the ship hits land or an underwater rock, sits on it and is not able to move); collision (includes both collision with other vessels/sea installations and sea-ice); fire; violence (meaning incidents towards persons and physical installations, whether environmentalists stopping activity, terrorists or pirates); other factors (including failure of the vessel such as construction or engine failure as the most frequent event). In this way, the accident type with the abbreviation "T-G" means the grounding of a tourist vessel, while "C-F" means a fire on a cargo vessel, and so on.

Table 3.1 Possible variation of accidents, depending on ship and event types

	Tourist	*Cargo*	*Fishing*
Grounding	T-G	C-G	F-G
Collision	T-C	C-C	F-C
Fire	T-F	C-F	F-F
Violence	T-V	C-V	F-V
Other (failure)	T-O	C-O	F-O

In the second step, using accident statistics, we define the probability of the specific type of events in the region (from theoretically possible to high-frequency), which we find at the place of event.

In the third step, we find the place of the event in the particular column, assuming the severity of consequences on the base of analysis and reports for similar accidents, size of the vessel, number of passengers, and fuel carried. It gives us the place of the event in the risk matrix (see Table 3.2, as an example for Svalbard's waters). Taking into consideration that the effects of the same event could be different for people and the environment, two separate matrices are introduced.

Qualitative expert evaluations on specific risk areas or defined situations of hazard and accident serve as the basis for the matrices. The estimate of consequences is based on case studies of the effects of real incidents in different parts of the world, illuminating accidents with different types of vessels. The analyses are also based on the results from exercises showing the capabilities of mitigating the negative effects of accidents in the Arctic (Gudmestad and Solberg, 2019) and knowledge on oil spill response (Eger 2010; Azzara and Rutherford, 2015; Wilkinson, Beegle-Krause, et al., 2017). For our assessment, we use the moderate scenario of the accidents as a base for the judgement on consequences.

Data sources include published reports on maritime activity in the Arctic and data published by emergency preparedness institutions on relevant issues in Norway, Iceland, Russia, and Greenland/Denmark. In addition, risk assessments have been discussed with industry specialists, government officials, researchers, navigators, and representatives from SAR-related authorities, organizations, and academic institutions. The qualitative data was collected and discussed in fora including researchers and emergency response professionals.

Data

Risk assessment calls for the collection and processing of relevant data. Several government and international initiatives have been taken in recent years to ensure the availability of data. Numerous research projects were

Table 3.2a Risk matrices for Svalbard waters

Risk of consequences for environment

	insignificant	minor	moderate	significant	serious
5 –Frequently					
4 Relatively frequently		F-G			
3 Periodically		F-C	T-G, T-O,		
2 Very rare		F-O, F-F	T-C,	C-O, C-C, T-F, C-F C-G,	
1 Theoretically possible			F-V, C-V, T-V		

Table 3.2b Risk matrices for Svalbard waters

Risk of consequences for people (passengers, crew)

	insignificant	minor	moderate	significant	serious
5 Frequently					
4 Relatively frequently		F-G			
3 Periodically		F-C	T-O	T-G	
2 Very rare		F-O	C-O, C-C, C-G	F-F, T-C	T-F, C-F
1 Theoretically possible				F-V, C-V	T-V

Note: Risk level: red: high; yellow: moderate; green: low.

funded to elaborate on the safety of shipping in the Arctic. Among these initiatives and projects are the WWF ArkGIS server (WWF, s.a.), the Canadian Arctic Shipping Risk Assessment System (Charlebois, Kubat, et al., 2017; Kubat, Charlebois, et al., 2017), the Baltic Marine Environment Protection Commission portal HELCOM (HELCOM, 2019) and the Search and Rescue in the High North (SARINOR) and Maritime Preparedness and International Partnership in the High North (MARPART) projects (Nord Universitet, 2018). The responsible safety authorities called for research on marine accidents, and these projects aimed to collect previous accident data and present them in a convenient way for analysis and decision-making.

Some official organizations provide open data access to statistics and analyses. For example, the Transportation Safety Board of Canada has data on its own website that was gathered from completed accident investigations from 1995, summarizing statistics and report (in total 506 reports since 1990 had been presented online by November 2019) (Transportation Safety Board of Canada, 2019). Some reports (a total of 81 since 2009 were available online by November 2019) can be found on the website of the Accidents Investigation Board of Norway (2019). Today, information about the new accidents is quickly distributed in media, including official publishers, the websites of information agencies, and on social networks. However, this information flow is not well ordered and systematized, and it is rather challenging to find the necessary data. For example, it can be challenging to find information about accidents in a certain area, at a certain time, and of a particular type. Thus, the information does not reach people who are interested in it or should be interested, according to their affiliation. That is why it is very important to make the data on marine accidents not only available but accessible, operable, and useful, put into the practical service of SAR and preparedness affairs.

Based on experience from collected data, government reports, and research devoted to maritime preparedness, we have created 'MarEmAr' – a web resource summarizing knowledge on handling marine emergencies, using ArcGIS software (ESRI, 2019).

'MarEmAr' stands for 'Marine Emergencies in the Arctic' and is a working title for our online geographical information system (GIS) solution, which presents case studies and important information about Arctic ship accidents. The main features are described in (Marchenko, 2019). 'MarEmAr' was created at the University Centre in Svalbard and is intended to condense experience gained from accidents; SAR, and salvage operations; and current available maritime accident investigation reports, in order to provide links to other resources that cover these accidents (e.g. web-pages, blogs, photo series, etc.). This online resource can be used for educating maritime professionals about various complications that lead to fatal accidents at sea.

Figures 3.6 shows two 'MarEmAr' windows zoomed in on the northern part of Svalbard and illustrate the difference between database and cases. Figure 3.6a demonstrates a database approach with many points on the map and standardized but very limited information, using the Norwegian Joint Rescue Coordination Center accident database (received in the frame of the MARPART project), listing all 862 events in 2000–18. Such information can be used for statistical analysis but gives limited knowledge. Figure 3.6b exposes case study layers: 'Svalbard Maritime Accidents', including 15 main cases (1989–2019) and 'Svalbard Exercises' with 11 exercises performed in 2014–19 to improve preparedness in the Arctic. Each point contains information on date and place, a short description of the event and a link to a detailed description. That allows users to gain data and knowledge in a desirable amount. The accidents can be selected and displayed by various characteristics: year, season, type, amount of

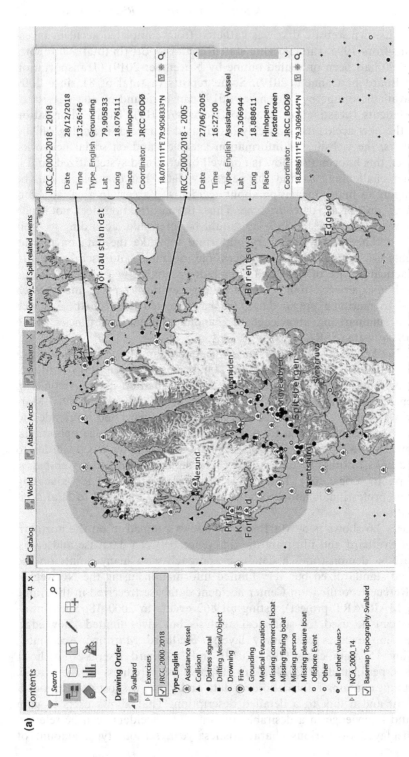

Figure 3.6a "MarEmAr" screenshot example: ArcGIS PRO view with "Svalbard Topography Basemap", and JRCC accident list layers switched on. The legend for the accident layer is on the left side, and the pop-up windows on the right side provide accident details.

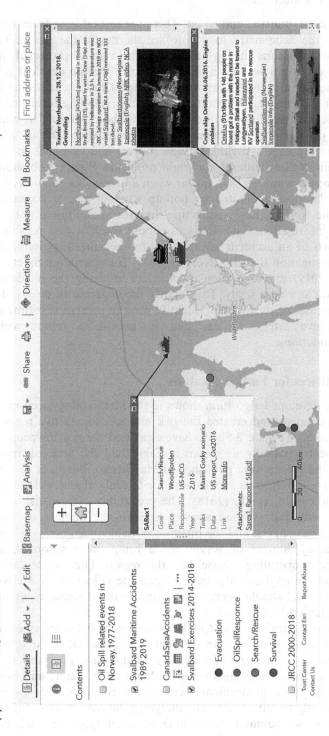

Figure 3.6b "MarEmAr" screenshot example: ArcGIS online view with two "Svalbard layers" turned on. Two events and one exercise pop-up windows are selected to show the description template. Blue words indicate links to web-pages with additional information (e.g. reports, articles, photo, video).

SAR involved, injuries, and so on. There are several base maps that can be obtained: the bathymetry map is to see the depth; the sea ice map is to reveal ice conditions during an accident; a map of protected areas and nature reserves can be used if there is concern that an accident could have environmental impacts. A map with settlements and rescue resources can be useful for planning SAR operations and understanding local preparedness capacities. The benefit of GIS technology is the possibility to update data and edit the map view on demand. The web location/state of 'MarEmAr' allows the use of a hyperlink to data sources and examines the details of the presented cases without copyright problems. Figure 3.6b shows examples of pop-up windows that appear when a user clicks on an event symbol. A short description, limited by pop-up window size, is accompanied by photos and 'INFO links', redirecting the user to additional webpages with accident information. The user can choose relevant and reliable sources. The reports of an accident investigation by specialized commissions (Accidents Investigation Board of Norway, 2019; Transportation Safety Board of Canada, 2019) are of particular value for studying the incident and its application as a case study. They clarify the sequence of events, determine the root causes of the accident, identify ways to prevent maritime accidents and improve safety at sea, and to publish a report with safety recommendations.

Risk Matrices for Further Analysis

Using the developed algorithm shown in the Methodology section, we made a risk assessment and created the risk matrices for all five regions of the Atlantic Arctic (Figure 3.5). We have estimated the risk for people and the environment separately, as shown in Table 3.2. The first risk matrices for Norway and Russia were published in Marchenko, Borch, et al. (2015) and Marchenko, Borch, et al., (2016) and for Greenland and Iceland in Marchenko, Andreassen, et al., (2018). The probability of high-risk event types increases with the growing activity level in the number of vessels, the number of passengers, and the presence of dangerous goods on each vessel. An increased number of vessels could bring more sailors with limited experience in operating a vessel into this region. The remoteness of Arctic routes and the cold climate make human life vulnerable if a crisis with a passenger vessel should occur, even with advanced rescue equipment (Solberg, Gudmestad, et al., 2017).

To assess the total risk and make comparisons between regions, we estimated the share of events with different risk levels, taking the total amount of chosen events to 100% (Figure 3.7). In this case, there are 15 different types of events: three defined types of ship (tourist, cargo, and fishing) and five defined types of accidents (grounding, collision, fire, violence, and others (mostly technical failure). Analyzing the risk matrices for the different regions, we counted the number of event types on each risk level.

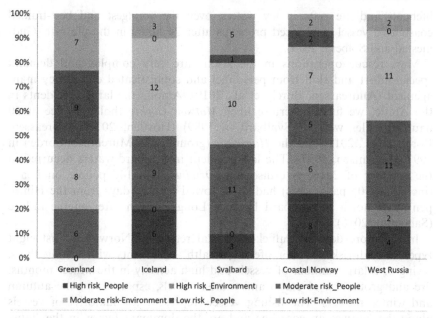

Figure 3.7 Share of events with different risk level in five regions of the Atlantic Arctic.

For example, for Russia, we emphasize four types of high-risk events for life and health – a collision of fishing vessels, and fires on cargo, fishing, and tourist vessels. For Svalbard, there are four other high-risk types for life and health events with tourist ships (collision, grounding, fire), as well as a fire on cargo vessels. Large cruise ships are the primary concern of the authorities on Svalbard. There are no other places in the world where cruise liners with 3,000 tourists on board sail up to 80°N. In the case of an accident, there are limited resources to save people in distress. The nearest ship to help could be several hours away. Two Super Puma helicopters based in Longyearbyen are not enough for mass evacuation. As a specific Svalbard exercise in November 2015 showed, two helicopters can evacuate 80 persons in seven hours operating at a distance of 50 km from Long-yearbyen (Svarstad, 2015). One can compare this number with average cruise vessels with 2,000 on board in the Magdalena fjord (the main tourist attraction) at 180 km distance.

Another case could be an accident involving expedition ships with 150 passengers on board in the Hinlopen Strait at the same distance, but with lower probability of another ship nearby. Hypothermia is the main issue in the case of large disasters in a very remote place in the Arctic. Owing to long distances, assistance cannot arrive quickly, and most likely in an emergency, people will need to wait for several hours. In an exercise testing survival in lifeboats and life rafts in icy waters, even the youngest and best-trained coastguard vessel crew faced problems after 24 hours in the life raft (Gudmestad and Solberg, 2019).

Mass rescue operations in the Arctic are very complex and demand special effort and skill from personnel and sophisticated emergency management (Andreassen, Borch, et al., 2019). Among the larger accidents in the Arctic, we find the cruise liner *Maksim Gorkiy* (holed by ice at 60 nautical miles west of Svalbard in 1989) (Hovden, 2012; Andreassen, Borch, et al., 2019) and the *Hanseatic* (grounded in Murchinsonfjorden in 1997) (Lorentsen, 1997). The last accident in Svalbard waters occurred in the summer of 2016. A cruise ship *Ortelius* with 146 people on board (including 105 passengers) had to be towed for two days from the Hinlopen strait north of Svalbard back to Longyearbyen after engine failure (Sabbatini, 2016).

In the more densely trafficked coastal region of Norway we list eight types of high-risk level for life and health – accidents with fishing ships, owing to a large number of vessels and high activity in the winter months; fire and grounding of cargo and tourist vessels, especially in the autumn and winter months. The fishing vessels represent the majority of vessels along the Norwegian coastline and are the dominant factor in the statistics of accidents at sea. This report included the number of injured and dead persons. For the first half of 2016, there were 123 persons injured and four persons who died at sea. Three out of four deaths were on a fishing vessel.

Until now, we have not experienced severe violent acts in the maritime Arctic. That is why all violent events are estimated as low risk, owing to limited probability. In general, regions with intensive traffic (coastal Norway, western Russia) could see a higher probability. However, at least in Norwegian coastal waters, there are more rescue resources and a higher level of preparedness, which makes it easier to find other vessels to help nearby ships in distress. The consequences for the environment are, however, more severe closer to land, where pollution recovery is both difficult and time-consuming. The estimates of pollution danger could increase, owing to more traffic with dangerous goods from Russian oil- and gasfields.

Figure 3.7 gives a visual representation of the proportion of events of varying risk degrees in the selected regions and of the general risk of marine activities. For the quantitative performance of the result, as a single number, we suggest calculating the Regional Risk Index (RRI). Table 3.3 shows the simple calculation. One of the possible approaches to calculating the RRI is to use a weight factor/co-efficient. For example, we can give a weight co-efficient of 1 to events of low risk, the weight of 2 to moderate risk events and the weight of 3 to high-risk events (first line in the Table 3.3 RRI calculation). Another approach is the percentage of normalization. We can use the following 'reference points':

Table 3.3 Regional Risk Index (RRI) key numbers and calculation. Aggregated (both for people and for environment) RRI is shown in bold. Particular RRI is shown in the parentheses (RRI for people/RRI for environment).

	Greenland	Iceland	Svalbard	Norway	Russia
	Number of events various risk degree (for people and environment)				
High risk	0	6	3	14	6
Moderate risk	14	17	21	12	22
Low risk	16	7	6	4	2
	Regional Risk Index (RRI), calculated in three different ways				
RRI with weight 1/2/3	**44** (21/23)	**59** (36/32)	**57** (32/25)	**69** (35/34)	**64** (34/30)
RRI with weight 1/3/5	**58** (27/31)	**88** (57/31)	**84** (49/35)	**109** (56/53)	**98** (53/45)
RRI with percentage approach	**48.4** (23.1/25.3)	**64.9** (39.7/25.3)	**62.7** (35.2/27.5)	**77** (38.5/27.6)	**70.4** (37.4/33)

Aggregated (both for people and for environment) RRI is shown in bold. Particular RRI is shown in parentheses (RRI for people/RRI for environment).

100% – All possible events have a high risk level; 0 – No events, no navigation in the area, zero risks; 66% – All possible events have a moderate risk level, 33% – All possible events have a low risk level (third line in the Table 3.3 RRI calculation). Through the first approach, we can vary the weighting factors, for example, using the 1/3/5 ratio to make the difference in the RRI more visible (second line in the Table 3.3 RRI calculation). Therefore, there is some variation in the results. The second approach seems to be more familiar, as people are accustomed to the degree of risk, to assess the development of the phenomenon as a percentage. However, all three approaches/techniques will give the same results if we convert the numbers in grades and reveal the risk degree (the colour in Table 3.3). The development of preparedness systems will stimulate the arrangement of different preparedness resources, organizations, and commanders that are involved in different cases. So, the calculation of a Regional Risk Index for different types of events can be requested. Taking it into consideration, we calculated separately a particular RRI for people and the environment and showed it in Table 3.3 in parentheses, and the aggregated RRI that is the sum of the two above is showed in bold in Table 3.3.

Conclusion

In this chapter, we have given an overview of various maritime activities in the Arctic and the risk of accidents with such large consequence that mitigation efforts from a broad range of resources are needed. The new regulations for icy waters, especially the Polar Code, reduce the probability of unprepared ship in the case of an accident. The Polar Code also demands survival equipment that could keep the personnel on board alive for five days if they have to evacuate. However, in order to develop an effective emergency response system, responsible agencies within the preparedness system should implement regional risk assessments. We suggest an assessment approach based on risk matrices providing a detailed Regional Risk Index and an online resource for data presentation.

Further validation of the risk assessment tools is important. Effective risk management decisions cannot be based exclusively on mapping ordered categorical ratings of frequency and severity. Optimal resource allocation could depend crucially on other quantitative and qualitative information. Therefore, distinguishing between the most urgent and least urgent risks in a setting with fast-changing conditions and the lack of incident statistics on the Arctic sea region is a challenging task. There is a need to reflect on the sudden appearance of 'black swan' incidents. To prepare for these rare, but dramatic events, qualitative judgements and worst-case scenario analyses are needed. Unexpected accidents can bring a combination of accidents, such as fire, injured and missing persons, and oil pollution. Thus, a minor accident in this region could fast escalate into a disaster.

In recent decades, emergency preparedness resources in the Arctic have been significantly strengthened by more available vessels and helicopters. However, the response time could be long, and the capacity is limited if major incidents occur. This situation calls for increased research efforts to learn more about how to reduce the probability of unwanted incidents. This analysis includes in-depth studies of modern vessel design and equipment, systems, and procedures, as well as the education and training of key personnel. We also need to look more closely into the preparedness capacities for both the private actor in the region as well as governments, regarding technology and personnel. We need to review the competences of both the vessel crew and the emergency response resources to deal with the challenges of Arctic waters. These investigations should include research on training and exercise schemes on less likely large-scale incidents. This demands efforts from a broad range of emergency response actors and cross-border support from other nations where institutional dimensions could represent an extra factor.

Limitations and Implications

Risk assessments in regions such as the Arctic should be based on a combination of quantitative and qualitative information. Categorizing severity can require inherently subjective judgements about consequences and decisions on how to aggregate multiple small events and fewer severe events. Therefore, risk matrices require a subjective interpretation. Qualitative risk matrices on emergency preparedness should be based on both the current statistics and estimates from experts from professional and research emergency preparedness institutions.

The following issues should be considered:

- Increasing traffic density, vessel capacity and size
- Oil and gas exploration and exploitation
- Efforts from international organizations, governments and industries to increase safety in Arctic waters and the differences in their approaches
- The availability of emergency capacities and their response time in different sea areas

For the categorization of consequences in the case of a lack of statistics in the Arctic region, the findings from the analysis of previous accidents, the conclusion of full-scale SAR, on-scene drills, and oil spill response exercises covering the different sea regions should be most fully used. That is why it is very important to make the data available and applicable. There is also a need to distinguish between the risk of severe consequences for the environment and humans. Consequences will always depend on different factors. Among them, preparedness and resource availability are the most important and some of the few factors that can be improved. We are not able to change the climate and make the harsh Arctic conditions milder, but we can enhance the emergency response resources and reorganize them where they are lacking, according to the regional risk assessment; we can use information about previous accidents and exercises and learn from the experience to improve the SAR and salvage procedures and make them more effective.

References

Abbassi, R., F. Khan, N. Khakzad, B. Vietch, and S. Ehlers (2017). Risk analysis of offshore transportation accident in Arctic waters. *The International Journal of Maritime Engineering*, January.

Accidents Investigation Board of Norway. (2019). Marine accidents. Retrieved from: www.aibn.no/Marine.

Andreassen, N., O. J. Borch, S. Kuznetsova, and S. Markov. (2019). *Emergency Management in Maritime Mass Rescue Operations: The Case of the High Arctic*. Sustainable Shipping in a Changing Arctic. L. P. Hildebrand, L. W. Brigham and T. M. Johansson, Springer Link: 359–381.

Arctic Council. (2011). *Agreement on cooperation on aeronautical and maritime search and rescue in the Arctic*. Nuuk: Arctic Council.

Azzara, A. and D. Rutherford. (2015). Air pollution from marine vessels in the U.S. High Arctic in 2025. ICCT- the international council on Clean Transportation (Working paper 1): 5.

Baksh, A.-A., R. Abbassi, V. Garaniya, and F. Khan (2018). Marine transportation risk assessment using Bayesian Network: Application to Arctic waters. *Ocean Engineering*, 159: 422–436.

Borch, O. J., N. Andreassen, N. Marchenko, V. Ingimundarson, H. Gunnarsdóttir, I. Iudin, S. Petrov, U. Jakobsen and B. Dali (2016a). Maritime activity in the High North – current and estimated level up to 2025.MARPART Project Report 1. Bodø, Nord University.

Borch, O. J., N. Andreassen, N. Marchenko, V. Ingimundarson, H. Gunnarsdóttir, U. Jakobsen, B. Kern, I. Iudin, S. Petrov, S. Markov and S. Kuznetsova. (2016b). Maritime activity and risk patterns in the High North: MARPART Project Report 2. Bodø, Nord University.

Charlebois, L., I. Kubat, P. Lamontagne, R. Burcher and D. Watson. (2017). Navigating in polar waters with CASRAS. *Journal of Ocean Technology*, 12(3): 43–52.

CHNL IO. (2019). Retrieved from: http://arctic-lio.com.

Cox, Jr, L. A. (2008). What`s wrong with Risk Matrices? *Risk Analysis*, 28(2): 497–512.

Crouhy, M., D. Galai. and R. Mark. (2006). *The Essentials of Risk Management*. New York: McGraw-Hill.

Dawson, J., L. Pizzolato, S. E. L. Howell, L. Copland and M. E. Johnston. (2018). Temporal and Spatial Patterns of Ship Traffic in the Canadian Arctic from 1990 to 2015 + Supplementary Appendix 1: Figs. S1–S7 (See Article Tools). *ARCTIC*, 71(1): 12.

DNV GL. (2014a). Analyse av sannsynligheten for akutt oljeslipp fra shipstrafikk ved Svalbard og Jan Mayen: 133.

DNV GL. (2014b). Miljørisoko knyttet til potensiell akuttoljeforurensning fra ship-strafikk i havområdene omkring Svalbard og Jan Mayen: 58.

DNV GL. (2019). The Arctic risk map. Retrieved from https://maps.dnvgl.com/arctic riskmap/.

Eger, K. M. (2010). *Effects of Oil Spills in Arctic Waters.*.

ESRI. (2019). Products. ArcGIS PRO and ArcGIS Online. Retrieved from www.esri. com/en-us/home.

Fiske, G. (2018). Ship traffic map. Unique Ship Visit (2009–2016). NASA Earth Observatory.

Fu, S., D. Zhang, J. Montewka, E. Zio and X. Yan. (2018). A quantitative approach for risk assessment of a ship stuck in ice in Arctic waters. *Safety Science*, 107: 145–154.

Gudmestad, O. T. and K. E. Solberg. (2019). Findings from two Arctic search and rescue exercises north of Spitzbergen. *Polar Geography*, 42(3): 160–175.

HELCOM. (2019). Map and data service. Retrieved from: http://maps.helcom.fi/web site/mapservice/.

Hovden, S. T. (2012). *Redningsdåden: om Maksim Gorkiy-havariet utenfor Svalbard i 1989*. Sandnes: Commentum.

International Maritime Organization. (2014). INTERNATIONAL CODE FOR SHIPS OPERATING IN POLAR WATERS (POLAR CODE). ANNEX. MEPC 68/21/Add.1: 54.

International Standardization Organization. (2018). ISO 31000 RISK MANAGEMENT. Retrieved from: www.iso.org/iso-31000-risk-management. html.

Khon, V. C., I. I. Mokhov, M. Latif, V. A. Semenov and W. Park (2010). Perspectives of Northern Sea Route and Northwest Passage in the twenty-first century. *Climatic Change*, 100(3–4): 757–768.

Kubat, I., L. Charlebois, R. Burcher, P. Lamontagne and D. Watson (2017). Canadian Arctic Shipping Risk Assessment System. 24th International Conference on Port and Ocean Engineering under Arctic Conditions. Busan, Republic of Korea.

Kystverket. (2019a). Havbase_Arktis. Retrieved from https://havbase.no/havbase_ arktis.

Kystverket. (2019b). Havebase. Retrieved from https://havbase.no/havbase.

Lentz, A. (2013). Risk of oil spills from groundings 2008 2009 (BRISK). 2008–2009. C. h. w. c. d. f. t. B. project. HELCOM.

Lorentsen, N. (1997). Fire døgn på skjæret. *Svalbardposten*, 28: 6–7.

Marchenko, N., N. Andreassen, O. J. Borch, S. Kuznetsova, V. Ingimundarson and U. Jakobsen (2018). Arctic Shipping and Risks: Emergency Categories and Response Capacities. *TransNav, International Journal on Marine Navigation and Safety of Sea Transportation*, 12(1): 107–114.

Marchenko, N. A. (2019). Marine Emergencies in the Arctic − GIS Online Resource for Preparedness, Response and Education. ISOPE − International Offshore and Polar Engineering Conference. Proceedings. Honolulu, USA.

Marchenko, N.A., O. J. Borch, S. V. Markov and N. Andreassen. (2015). Maritime activity in the High North – the range of unwanted incidents and risk patterns. The 23rd International Conference on Port and Ocean Eng. under Arctic Conditions (POAC 2015), Trondheim.

Marchenko, N.A., O. J. Borch, S. V. Markov and N. Andreassen (2016). Maritime safety in the High North – risk and preparedness. 26th International Ocean and Polar Engineers conference (ISOPE-2016), Rhodes, Greece.

Nord Universitet. (2018). MARPART -Maritime preparedness and International Collaboration in the High North. Retrieved from: www.marpart.no.

Pravetoni, R. (2010). Towns and industrial activities in the Arctic. Protecting Arctic Biodiversity.

Sabbatini, M. (2016). Ship out of luck: Governor tows vessel with 146 people back to Longyearbyen after engine failure. *Ice People*, Longyearbyen.

Solberg, K. E., O. T. Gudmestad and E. Skjærseth (2017). SARex. Surviving a maritime incident in cold climate conditions. University of Stavanger. 69: 269.

Svarstad, S. M. (2015). Øvelse Svalbard. National helseøvelse 2014. Største i historien på Svalbard.

Trbojevic, V. M. C., B. J. Carr. (2000). Risk based methodology for safety improvements in ports. *Journal of Hazardous Materials*, 71: 467–480.

Transportation Safety Board of Canada. (2019). Marine transportation safety investigations and reports. Retrieved from: www.bst-tsb.gc.ca/eng/rapports-reports/marine/ index.html.

UN Office for Disaster Risk Reduction. (2009–17). Terminology on Disaster Risk Reduction.

Wilkinson, J., C.J. Beegle-Krause, K.-U. Evers, N. Hughes, A. Lewis, M. Reed and P. Wadhams (2017). Oil spill response capabilities and technologies for ice-covered Arctic marine waters: A review of recent developments and established practices. *Ambio*, 46(Supplement 3): 423–441.

WWF. (s.a.). ArkGIS (Arctic Geographical Information System). Retrieved from: http://wwfarcticmaps.org/.

4 Search and Rescue Capabilities in the Arctic

Is the High North Prepared at an Adequate Level?

Dimitrios Dalaklis and Megan Drewniak

Introduction

There are several efforts to categorize the Arctic regions. Among others, we find the term 'High North'[1]. The High North could be described as the areas around the North Pole. Even though ice coverage is diminishing, the polar ice cap still dominates these high latitudes. The ice areas consist mainly of ice floes and pack ice, 7–10 feet (2–3 m) thick. For the current analysis, the Arctic is considered to comprise the Arctic Ocean and the territories within the Arctic Circle. We adopt the definition used by the five states (the '**Arctic-Five**'), namely Canada, the USA, Norway, Denmark (through Greenland), and the Russian Federation. Their geographic position is creating important obligations in terms of safety. They are located adjacent to the emerging maritime corridors and bear the main responsibility to render assistance in cases of need (Dalaklis and Drewniak, 2018). The Arctic landscape is being transformed at an unprecedented speed, as evidenced by the scientifically recorded decline of ice coverage in the region. This decline creates significant new business opportunities. The shipping industry is directly affected by the many touristic endeavours and ambitious energy resource exploration projects already under way[2], as well as the opening of the so-called 'Arctic Passages' (Dalaklis et al., 2018a).

Theory and Methodology

The complex operating environment for seagoing vessels engaged in trade or tourist and fishing activities in the High North means that ensuring safety should always be a high priority in the regulatory agendas of concerned nations. These factors should strongly influence their decisions to invest in further supporting infrastructures for ships operating in the region. Having previously examined the multi-faceted geopolitical dimensions of the Arctic (Dalaklis and Baxevani, 2016; Seker and Dalaklis, 2016; Drewniak et al., 2018, Dalaklis et al., 2018b), the challenges of an increased level of maritime traffic in that region should be brought into the discussion, along with specific associated safety implications. In this chapter, we first provide a brief analysis

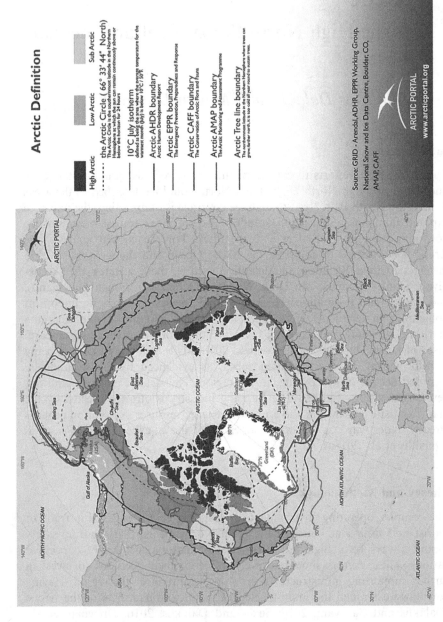

Figure 4.1 Summary of Different Approaches towards an "Arctic Definition" (Arctic Portal, available: https://arcticportal.org/maps-?arctic-definitions)

of the primary maritime routes into the broader region. Furthermore, as human presence and operations are clearly expected to intensify within the Arctic, it is of utmost importance to evaluate the associated safety risks and (if needed) introduce the necessary mitigation measures. Therefore, the infrastructure currently available to support the conduct of search and rescue (SAR) operations will be qualitatively evaluated. The international legal framework associated with the conduct of SAR will be summarized to achieve this aim, followed by the examination of how the Agreement on Co-operation on Aeronautical and Maritime Search and Rescue in the Arctic is operationalized in the field by the various states involved.

As discussed in Chapter 3, 'risk' most commonly relates to the probability of uncertain future events as the product of the 'likelihood' and 'impact' of a potential event (Cline, 2004). Statistically, the level of downside risk can be calculated as the product of the probability that harm occurs (e.g., that an accident happens) multiplied by the severity of that harm (i.e., the average amount of harm or more conservatively the maximum credible amount of harm). Factors that can influence the risk level are shown in Figure 4.2 (Borch, 2019). Borch puts forward the notion that the risk can be calculated as the product of the probability of an accident, the resulting consequences, and of course the resulting vulnerability (Risk = probability of an accident x consequence x vulnerability). Therefore, in order to bring the risk of Arctic operations down, an evaluation (and improvement if needed) of the associated SAR infrastructure is essential, considering that there will be a positive influence in both the 'consequence' and 'vulnerability' factors.

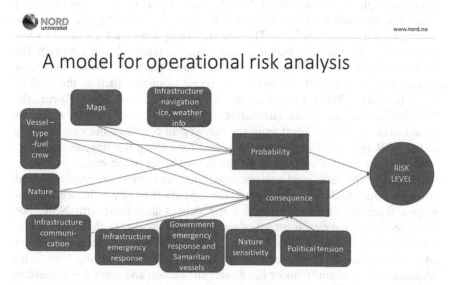

Figure 4.2 "Managing Risk" in the Arctic (Borch, 2019)

There are significant variances in the traffic volume from year to year, and hindrances persist. The availability of icebreakers needed to keep the maritime corridors open and provide an escort to transiting vessels stands out clearly as a barrier (DeWitz et al., 2015; Dalaklis and Drewniak, 2017; Drewniak et al., 2017; Drewniak and Dalaklis, 2018; Dalaklis et al., 2018b). The introduction of the International Code for Ships Operating in Polar Waters (Polar Code) has already had a positive influence on the safety level of those ships that operate in the still-harsh Arctic environment[3]. However, apart from the frozen and hostile environment, the vast distances that must be travelled in time of need still represent significant challenges for the responding organizations and the people involved with SAR in the Arctic context.

Data and Analyses

Shipping is an industry strongly interwoven with the environment, and as such, it can be directly affected by the latest developments in the Arctic (Dalaklis, 2017). The retreat of ice in the region is opening new areas/routes for navigation. According to the latest scientific estimates, the days when navigation is possible are expected to follow an increasing trend: from around 70 days (currently) up to 125 in 2050 and as many as 160 in 2100 (Cariou and Faury, 2015). Significant discussions have emerged on the potential profits/financial benefits that the use of the 'Arctic Passages' could yield. Regarding Arctic navigation, it is necessary to highlight that the greatest interest is directed towards two different routes (Figure 4.3): a) Northwest Passage (NWP); and b) Northern Sea Route (NSR), which are both intercontinental maritime connectivity alternatives.

The importance of the NSR lies mainly in the fact that it can connect the Asian and European markets through the eastern part of the Arctic Ocean. It offers an alternative to the Suez Canal route by reducing the respective voyage by approximately 40%. This alternative option could also lead to less pressure on the main transcontinental route currently in use (through the Malacca Strait–Indian Ocean–Suez Canal) and allow for the avoidance of regions that are currently considered as prone to piracy, such as the Gulf of Aden (Dalaklis, 2012). For states like China, Japan, and South Korea, the NSR could also provide an alternative to using the extremely busy and congested Malacca Strait. Apart from the Norwegian coastline, the vast majority of the NSR (about 90% of the entire route) runs along the Russian coastline, from Novaya Zemlya in the west to the Bering Strait in the east.

Because of this fact, legal complications arise, as there is the regulation that puts this route under Russian jurisdiction. Apart from the influence of geography, there are also certain legal complications associated with NSR. For example, the formal establishment of the NSR was decided in the Soviet era by the Council of People's Commissars of the Union of Soviet Socialist Republics on 17 December 1932. This same regulation also dictates that the NSR is an administered, legal entity under full Soviet jurisdiction and control – a position maintained today by Russia. Furthermore, this law creates the necessity to pay

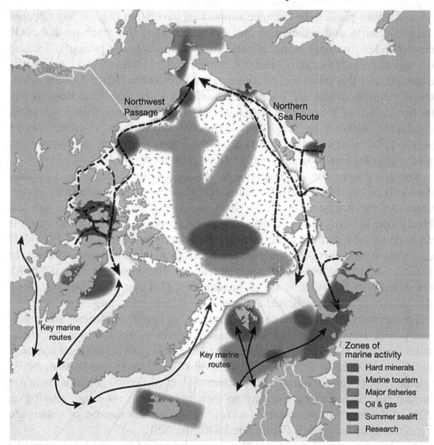

Figure 4.3 "Arctic Passages" and Zones of Activity in the Arctic (Source: Hugo
   Ahlenius, GRID-Arendal & CAFF, available online: https://www.grida.no/
   resources/6274)

a fee for the obligatory vessel escort by Russian operated icebreakers and
mandates the acquisition of a permit in order to cross these waters[4]. These
obstacles limit the attractiveness of the NSR in terms of introducing 'additional
costs' and, as argued by third-party states, pose a de facto limitation on the
freedom of navigation. On the positive side, the Russian state has recently significantly simplified the bureaucratic procedure for the issuance of a crossing
permit, which used to be a major hindrance (Dalaklis and Baxevani, 2017).

   Cargo traffic volume in the NSR experiences significant fluctuations, considering that it can be influenced by several economic, political, seasonal, and
geopolitical factors[5]. At this point, the primary cargoes being shipped along
the NSR include energy resources (fuel), timber, equipment and various
commodities, including large volumes of raw materials. Aiming to promote
the use of the NSR, the Russian state has invested heavily in maritime

infrastructure projects to enhance support of navigation (e.g., it has updated nautical charts and established and operationalized the much-needed SAR centres)[6]. For the time being, general cargo shipments are primarily focused on regional needs (terminal and not transit traffic [Lasserre, 2012]), and it is expected that this trend will continue in the foreseeable future. Locally oriented shipping activities are usually conducted by ships under the Russian flag (with old, but still capable ice-strengthened vessels; these activities are often conducted without an escorting icebreaker during the so-called 'warm season'). In addition, the direction of the cargo is rather unbalanced, as return voyages generally carry only ballast (Humpert, 2014).

The NWP links the Atlantic with the Pacific Ocean, through the Canadian Archipelago (see Figure 4.3). Currently, this passage is a very interesting opportunity for the tourism industry (in 2016 the *Crystal Serenity* became the first luxury liner to cross with more than a thousand people on board; this activity was repeated in 2017, but not since then) (Dalaklis and Drewniak, 2018). Individual cargo ships have also used this passage, although not in significant numbers. In general, cargo ships would have fewer limitations regarding their draft and width compared with the Panama Canal prerequisites. However, the geography along this passage is very complicated and rather troublesome for navigation.

Furthermore, drifting ice can block access to specific locations and entry/exit points, leaving only narrow or shallow corridors open, which are unsuitable for large vessels. In addition, Canada's small icebreaking fleet makes it nearly impossible to deliver the same level of services as the Russian NSR-related fleet (Drewniak et al., 2017; Drewniak et al., 2018). Finally, there are hardly any points along the way to pick up or drop off cargo (Rodrigue, 2016). Because of these obstacles, a significantly smaller number of vessels are expected to cross the NWP compared with the NSR.

Another passage of interest is the Arctic Bridge (AB), which connects the Canadian port of Churchill, the Norwegian port of Narvik, and the Russian port of Murmansk. For this option, it must be noted that as Greenlandic ice melts, the number of icebergs that can be encountered for vessels transiting the AB will be increased. Discussion of the Arctic routes also involves the Central Arctic Route or Transpolar Route, which provides a 'direct line' of passage across the Arctic Ocean. For the moment, the few voyages completed via this passage have been exploratory (scientific expeditions, the testing of travelling times, and the use of equipment in extreme weather conditions) or for power projection reasons. For this route to be considered commercially viable, numerous changes must take place regarding infrastructure, communications, emergency response, ice-capable fleets and, above all, an extreme continuation of sea-ice decline (Dalaklis and Baxevani, 2017). Before moving in a different direction, it is necessary to highlight again that projections for increased shipping traffic along both the NSR and NWP are encouraged by substantial reductions in the distance compared with the Suez and Panama Canal routes. Taking advantage of these routes has the potential for large cost savings because of reduced fuel consumption and increased trip frequency (Lasserre, 2014).

The SAR services of the Arctic depend heavily on ships. For the most part, vessels of different sizes and capabilities engage in defence/law enforcement duties, trade and cruising/leisure activities, or fishing. To effectively assist vessels and persons in distress at sea, aircraft and helicopters or even unmanned aerial vehicles can also contribute significantly to that effort. The 1982 United Nations Convention on the Law of the Sea (UNCLOS Convention), in Article 98 (1), provides that:

> Every State shall require the master of a ship flying its flag, in so far as he can do so without serious danger to the ship, the crew or the passengers:
> (a) to render assistance to any person found at sea in danger of being lost;
> (b) to proceed with all possible speed to the rescue of persons in distress, if informed of their need of assistance, in so far as such action may reasonably be expected of him.

The same Article of UNCLOS lays down the general obligation for coastal states to arrange a SAR service. Itt is clearly stated that every coastal State Party must:

> ...promote the establishment, operation and maintenance of an adequate and effective search and rescue service regarding safety on and over the sea and, where circumstances so require, by way of mutual regional arrangements co-operate with neighbouring States for this purpose.

The obligation of ships to assist other vessels in a distress situation at sea is enshrined both in tradition and in international treaties, with the International Convention for the Safety of Life at Sea (SOLAS Convention, 1974) clearly standing out. It is therefore not a coincidence that Chapter V, Regulation 7 of SOLAS requires State Parties:

> ... to ensure that necessary arrangements are made for distress communication and co-ordination in their area of responsibility and for the rescue of persons in distress at sea around its coasts. These arrangements shall include the establishment, operation and maintenance of such search and rescue facilities as are deemed practicable and necessary ....

Rescuing persons from ships or boats in distress, apart from its obvious humanitarian responsibility dimension, is also very clearly established as a requirement under SOLAS Chapter V, Regulation 33, which states (in part) that:

> The master of a ship at sea which is in a position to be able to provide assistance on receiving information from any source that persons are in distress at sea, is bound to proceed with all speed to their assistance if possible informing them or the search and rescue service that the ship is

doing so. This obligation to provide assistance applies regardless of the nationality or status of such persons or the circumstances in which they are found.

The legal infrastructure/systems that have already been set up under the auspices of the International Maritime Organization[7] (IMO) – in close collaboration with the International Civil Aviation Organization[8] (ICAO) – and the exploitation of various modern technology applications facilitate today the swift and easily coordinated conduct of a (search and) rescue operation. The overall framework of SAR corresponds to a set of duties that include the search and rescue of persons in distress at sea, provision of emergency medical services for them and conduct of radio communications related to an emergency phase. Other duties regarded as part of maritime SAR include the provision of telemedical assistance services for vessels, maritime assistance services, the use of certain emergency signalling devices and emergency medical services at sea. To clarify further, the term 'search' describes an operation, normally coordinated by a rescue coordination centre or rescue sub-centre, using available personnel and facilities to locate persons in distress. The term 'rescue' describes an operation to retrieve persons in distress, provide for their initial medical or other needs, and deliver them to a place of safety.

Jointly published by IMO and ICAO, the three-volume International Aeronautical and Maritime Search and Rescue (IAMSAR) manual provides guidelines for a standard aviation and maritime approach to organizing and providing SAR services. The manual harmonizes aeronautical and maritime SAR organization, procedures, equipment, and terminology internationally; Nations 'implement accordingly' the Global SAR system developed by IMO and ICAO. Each volume can be used as a standalone document or, in conjunction with the other two volumes, as a means to attain a full view of the SAR system. The IAMSAR manual is divided into three volumes: a) Volume I, Organization and Management, discusses the global SAR system concept, establishment and improvement of national and regional SAR systems and co-operation with neighbouring states to provide effective and economical SAR services; b) Volume II, Mission Co-ordination, assists personnel who plan and co-ordinate SAR operations and exercises; and c) Volume III, Mobile Facilities, is intended to be carried on board rescue units, aircraft, and vessels[9] to help with the performance of a search, rescue, or on-scene co-ordinator function, as well as with aspects of SAR that pertain to their own emergencies.

Those in need anywhere on the high seas can immediately 'call for help'. Distress signals can be rapidly transmitted by satellite and terrestrial communication systems both to search and rescue authorities ashore and to ships in the immediate vicinity, therefore facilitating the timely delivery of SAR services to those in danger.

It is important to emphasize that the International Convention on Maritime Search and Rescue (SAR Convention, 1979) is a very influential international agreement concerning the important issue of maritime SAR services. The Convention was adopted at a conference in Hamburg in 1979 (it was adopted on 27 April 1979 and entered into force on 22 June 1985). It deals with the need to develop an international SAR plan, so that, no matter where an accident occurs, the rescue of persons in distress at sea will be co-ordinated by an SAR organization and, when necessary, by co-operation between neighbouring SAR organizations. Following the adoption of the SAR Convention, IMO's Maritime Safety Committee divided the world's oceans into 13 SAR areas, in each of which the countries concerned all have delimited SAR regions for which they are responsible. For the Arctic Region, there is a dedicated 'Arctic Search and Rescue Agreement' (its formal title being The Agreement on Cooperation on Aeronautical and Maritime Search and Rescue in the Arctic), which is an international treaty concluded among the member states of the Arctic Council (Canada, Denmark, Finland, Iceland, Norway, Russia, Sweden, and the USA) on 12 May 2011 in Nuuk, Greenland. The government of Canada is the depositary for that treaty, and it entered into force on 19 January 2013, after each of the eight signatory states had ratified it.

The Arctic Council (AC) is an institution that is prevalent in Arctic affairs. It was established as an intergovernmental forum in 1996, and it has eight member states: the 'Arctic Five', plus Finland, Iceland, and Sweden[10]. Its goal is to facilitate communication and co-operation among Arctic states in dealing with environmental protection, sustainable development initiatives, SAR operations, and a variety of other regional issues. Despite poor performance reviews from its critics, it has made important contribution progress in the development of doctrine, which will help to establish a co-ordinated framework for Arctic navigational safety and environmental protection. Specifically, the Agreement on Co-operation on Aeronautical and Maritime Search and Rescue and the Agreement on Co-operation on Marine Oil Pollution Preparedness and Response in the Arctic, are key pieces of legislation which will help to guide and unify Arctic member states' approach in dealing with a wide range of critical regional issues (Drewniak and Dalaklis, 2018)[11].

With more traffic being expected in the Arctic region, the need for a sound SAR framework is essential. This previously mentioned agreement co-ordinates life-saving international maritime and aeronautical SAR coverage and response among the Arctic States across an area of approximately 13 million square miles (Figure 4.4). Furthermore, through this framework of co-operation, competent authorities for each of the eight member states and the agencies responsible for carrying out SAR activities were established. Lists of the necessary rescue co-ordination centres (RCC) by name and location for each member state are also included (International Security Advisory Board, 2016). This agreement is not prescriptive in how the states concerned must physically implement a response position and is, therefore, a bit lacking. Arctic nations must further build up the infrastructure supporting the safe conduct of navigation in both the NSR and the NWP; developing and then

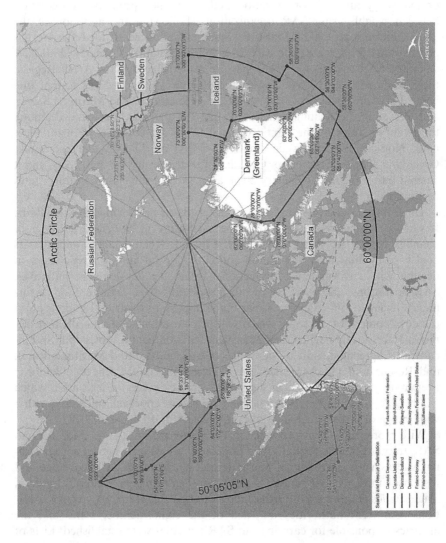

Figure 4.4 Arctic Search and Rescue Agreement-Areas of Application (Arctic Portal, available online: https://arcticportal.org/)

testing contingency plans of an international nature to prepare for the possibility of large-scale maritime accidents that require SAR and oil spill response in the harsh Arctic conditions is a priority measure.

Russia has responsibility for the SAR mission along the NSR. To address concerns associated with the safe transit of vessels along this passage, Russia has invested 910 million rubles (US$30.1 million) in the creation of ten SAR centres (Murmansk, Dickson, Arkhangelsk, Naryan-Mar, Vorkuta, Nadym, Tiksi, Pevek, Provideniya, and Anadyr) (Drewniak and Dalaklis, 2018). Oversight of SAR and oil spill response along the NSR is co-ordinated by a maritime rescue co-ordination centre (MRCC) and two maritime rescue sub-centers (MRSC) (Figure 4.5). The MRCC is located in Dickson, and the MRSCs are located in Tiksi and Pevek. The presence of icebreaking vessels in Dickson is year-round. Designated forwarding operational locations (FOL) offer additional support between July and October and involves Tiksy, Pevek, and Provideniya. The NSR Information Office also lists five icebreakers (NS *Yaygach*, NS *Yamal*, NS *50 Let Pobedy*, NS T*aimyr*, and *Admiral Makarov*), two of which also have on-board diving equipment and oil spill response equipment, capable of providing support for vessels in the NSR.

Despite these proactive measures to lessen the risk associated with travelling this route with an increased SAR presence, the centres/sub-centres and more important assets involved with SAR tasks such as vessels, boats, and helicopters are still separated by rather vast distances, and the response time could easily be inadequate to prevent fatalities in an emergency. West of Russia, the Joint Rescue Co-ordination Center of Northern Norway is a rescue co-ordination centre located in Bodø, which has responsibility for co-ordinating SAR operations in the northern part of Norway, north of the 65th parallel (Government of Norway, 2017). Norway and Russia have a bilateral SAR agreement for the Barents Sea, of which the content is renewed and practiced annually (Dalaklis, 2019).

The land area along the NWP is extremely underdeveloped, especially around the waterways of the Canadian Arctic, with no port of significant size. Port facilities along the North American Arctic coast west of the passages are equally negligible, and the closest well-developed port is Nuuk (on the west coast of Greenland); this is the largest and most significant port of the NWP. The Canada Shipping Act of 2001 governs shipping in the Canadian Arctic. The Act established a framework for SAR and communication operations in Canadian Arctic waters. It included the following services: monitoring international marine radio distress frequencies, broadcasting ice and marine weather information and notices concerning hazards to navigation, and screening ships entering Arctic waters to enhance safety and prevent pollution. A party to the Arctic 2011 SAR Agreement, The Canadian Ministry of National Defence is designated as the Competent Authority in the Arctic Council's 2011 SAR Agreement, and both the Canadian Coast Guard and Canadian Forces are designed as SAR agencies. Furthermore, the designated

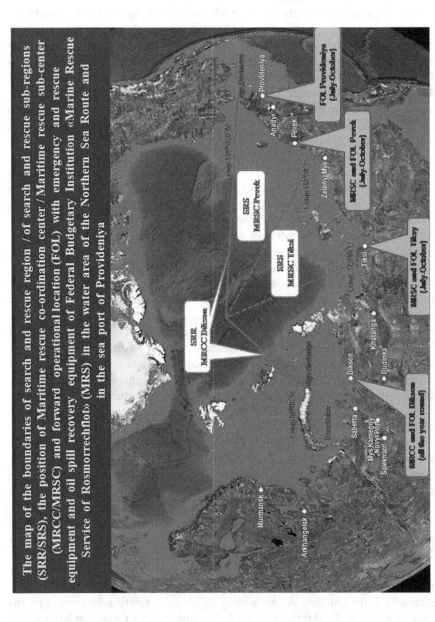

The map of the boundaries of search and rescue region / of search and rescue sub-regions (SRR/SRS), the position of Maritime rescue co-ordination center / Maritime rescue sub-center (MRCC/MRSC) and forward operational location (FOL) with emergency and rescue equipment and oil spill recovery equipment of Federal Budgetary Institution «Marine Rescue Service of Rosmorrechflot» (MRS) in the water area of the Northern Sea Route and in the sea port of Provideniya

Figure 4.5 Rescue Coordination Centers/Areas of Responsibilities along the NSR (Source: Northern Sea Route Administration, available online: http://www.nsra.ru/en/pso.html)

rescue co-ordination centre for Canada is in Trenton, and the respective SAR area of responsibility is provided via Figure 4.4 (Dalaklis et al., 2018b).

The waters around Alaska fall under US jurisdiction and are governed by the US Coast Guard (USCG) which is fundamentally responsible for maritime safety, national defence, maritime security, maritime mobility, the protection of natural resources, and ice operations. As the lead agency for maritime SAR, the USCG maintains a coastal network of boat stations, aircraft, communication systems, and a command-and-control network to respond to those in peril at sea. The USCG is designated as the Competent Authority in the Arctic Council's 2011 SAR Agreement and both the USCG and US Department of Defence are designated in that legal instrument as SAR agencies. The designated joint rescue coordination centre for the USA is located in Juneau, Alaska, and an Aviation Rescue Coordination Centre is located in Elmendorf, near Anchorage, Alaska[12]. Given the instrumental role that the USCG plays in maritime operations within Arctic waters, there are multiple capital investment needs in infrastructure to include icebreaking capabilities and the shore-based infrastructure to support both SAR and icebreaker operations. Specifically, the USCG has an inadequate force structure to meet SAR contingencies, and investment in constructing ice-strengthened patrol ships and aircraft for extreme climates is needed (Deboer, 2017; Dalaklis and Drewniak, 2019).

Conclusion

In recent years, the Arctic's ice coverage has been on a steady decline. Recognizing the rise of an opportunity, a fairly significant number of business entities already operate within (or have considered exploiting) the High North. The focus is upon previously untapped resources, such as precious minerals and large quantities of oil and gas. Tourism and fishing activities are clearly intensifying, with various endeavours of maritime transport also being put forward. The Arctic is now viewed as a promising field for economic activities and is considered as a potential connecting corridor between Asia and Europe/North America (and vice-versa). It is the shipping industry, however, that is presented with the most daunting challenges and perhaps stands to reap the most significant benefits. Not since the opening of the Suez and Panama Canals has this industry seen such transformative changes with the potential to introduce a new era to shipping.

While the NWP and NSR have both attracted attention from much of the academic and business world, their contribution to global maritime traffic volume is still minimal. Although using the 'Arctic Passages' could decrease the total number of nautical miles that a vessel needs to travel to reach its desired destination, a precise determination regarding the extent of cost savings to the maritime industry is so far inconclusive, and more research is needed. There are increased dangers of operating in the Arctic, including harsh environmental conditions and especially extreme weather conditions.

Among the challengers are polar darkness; ice floes, long distances without adequate support; a lack of emergency response capabilities to cover the whole area; and challenging shipbuilding and equipment requirements. Any of these alone is very difficult to overcome and could nullify the advantage of the voyage distances that these 'Passages' can provide.

The existence of adequate SAR centres and their performance are points of concern. More surveillance and communication systems should be considered. Nautical charts are currently being updated, and a dedicated service to monitor and inform transiting maritime traffic on obstructing ice floes is crucial. Despite the endless economic, scientific, and technological possibilities that the Arctic can offer, the maritime community must understand that further development in this region must be done with the utmost consideration for safety and environmental stewardship. It is pivotal that sustainability constitutes an intrinsic aspect of regional development. The lack of transport infrastructure, the necessity for specialized training and equipment, and suboptimal emergency response capabilities are only a few of the variables that severely hinder the integration of the 'Arctic Passages' as viable and lucrative global trading routes. For the cruise industry, any SAR event associated with a large cruise ship, or even a smaller 'expedition' or 'adventure' vessel, will involve an extended number of people that clearly exceeds the capabilities of the current infrastructure that is in place.

The existing number of ice-classed SAR vessels operating in the Arctic cannot satisfactorily cover the vast areas. Survival time in these waters is well below the respective time for warmer areas. Unless there is extraction by helicopter, or there is a ship in very close proximity, casualties could be a consequence. The emergency preparedness systems and response capabilities available to respond to human and environmental catastrophes must be further developed in order to bridge gaps. Further research and intensive capacity-building efforts will be needed to alleviate the weaknesses of the emergency response system. Live training drills and exercises to further evaluate the current regulations and identify points for improvement should be the priority.

Notes

1 The terms 'Arctic' and 'High North' are used interchangeably.
2 Tourist activities are clearly on the rise over the last couple of years; the number of ships operating in the Arctic will certainly increase in the future, especially by considering activities related to the 'Yamal LNG' (an integrated project encompassing natural gas production, liquefaction, and shipping) and the upcoming 'Arctic LNG-2' projects.
3 With conditions for ships in the Arctic still remaining harsh and perilous, the International Maritime Organization (IMO) has already taken a forward-leaning approach toward the enhancement of maritime safety, via the introduction of the Polar Code on 1 January 2017. The IMO has adopted the Polar Code and related amendments to make it mandatory under both the International Convention for the Safety of Life at Sea (SOLAS) and the International Convention for the

Prevention of Pollution from Ships (MARPOL). The Polar Code and SOLAS amendments were adopted during the 94th session of IMO's Maritime Safety Committee in November 2014; the environmental provisions and MARPOL amendments were adopted in May 2015.

4 Another relevant maritime traffic regulation to note is the '2012 Federal Law No. 132', adopted by the Russian State (Amendments to Certain Legislative Acts of the Russian Federation Related to the Governmental Regulation of Merchant Shipping in the Water Area of the Northern Sea Route). A broad 'definition' of the NSR is provided there; the specific legal text also describes the prerequisites for crossing (permit and fee).

5 Current traffic levels in the region can be explained by the overall structure of the shipping sector. Containerships are reliant on predictability and accuracy (Corbett and Winebrake, 2008), conditions that the Arctic Passages can by no means provide (Schøyen and Bråthen, 2011). For this reason, dry bulk and vessels engaged in the transport of commodities needed locally are the ones most frequently involved with operations in the Arctic (Rodrigue, 2013). Cruising and other touristic activities are also now conducted more frequently; for the near future, the influence of energy resources and especially shipments of liquefied natural gas should also be factored into this equation.

6 Currently, there are 17 ports along the NSR, and icebreaking services grant uninterrupted access to the majority of them (NSR Administration, 2016; Dalaklis and Baxevani, 2017). The stated goal of the Russian state is to promote the use of NSR to assist Russia in re-gaining the highest levels of Soviet-era global economic influence and power. To meet this objective and make this route more appealing to the shipping industry, it has invested significantly in the region to build port infrastructure, acquire the necessary capacity to provide additional icebreaking capabilities, enhance aids to navigation, and efforts to provide better access to inland customers are already under way.

7 IMO is the United Nations (UN) specialized agency with responsibility for the safety and security of shipping and the prevention of marine and atmospheric pollution by ships.

8 ICAO is a UN specialized agency that manages the administration and governance of the Convention on International Civil Aviation (Chicago Convention). Prompted by the need to rapidly locate and rescue survivors of aircraft accidents, a set of internationally agreed Standards-Recommended Practices has been incorporated in ICAO's Annex 12 – Search and Rescue (SAR). The Annex, which is complemented by a three-part Search and Rescue Manual dealing with SAR organization, management and procedures, sets forth the provisions for the establishment, maintenance and operation of search and rescue services by the ICAO Contracting States in their territories and over the high seas.

9 In order to ensure the quality of SAR services, SOLAS Chapter V (Safety of Navigation) requires ships to carry an up-to-date copy of Volume III of the IAMSAR Manual. As a result, 'SOLAS-type' vessels should be considered as already equipped with the necessary skills and tools to immediately get involved with SAR operations, although the previous level of familiarity of shipmaster/crew involved with this type of operations should also be viewed as an important element towards a successful outcome. IMO has already recognised the need for expanding the knowledge of those who may be designated to perform the duties and responsibilities of a Search and Rescue On-Scene Co-ordinator for an SAR incident, as defined in IAMSAR manual vol. III. There is a designated Model Course (3.15) assisting States in meeting their SAR obligations under SOLAS and the International Convention on Maritime Search and Rescue (SAR), 1979 as amended.

10 More details and a description of work currently underway can be found at http s://arctic-council.org/index.php/en.

11 To provide some direction on the issue of response to potential oil spills in the Arctic region, the AC facilitated the Agreement on Co-operation on Marine Oil Pollution Preparedness and Response in the Arctic, which was signed in 2013. Its objective to strengthen co-operation, co-ordination, and mutual assistance among the parties on oil pollution preparedness and response in the Arctic in order to protect the marine environment from pollution by oil.

12 The SAR area of responsibility for the U.S. is also designated in Figure 4.4.

References

Borch, O. J. (2019). Maritime Preparedness in Arctic Waters. Association of Arctic Expedition Cruise Operators' 16th Arctic Cruise Conference, Oslo-Norway.

Cariou, P. and Faury, O. (2015). Relevance of the Northern Sea Route (NSR) for bulk shipping. *Transportation Research Part A*, Vol. 78, pp. 337–346.

Cline, P. B. (2004). The Merging of Risk Analysis and Adventure Education. *Wilderness Risk Management*, Vol. 5, No. 1, pp. 43–45.

Corbett, J. J. and Winebrake, J. (2008). The impacts of globalisation on international maritime transport activity: past trends and future perspectives. Energy and Environmental Research Associates, the United States, OECD. Global Forum on Transport and Environment in a Globalising World, Guadalajara, Mexico.

Dalaklis, D. (2012). Piracy in the Horn of Africa: Some good news, but a lot of work has still to be done. *Maritime Security Review-MSR InDepth*, No. 9, pp. 1–7.

Dalaklis, D. and Baxevani, E. (2016). Arctic in the global warming phenomenon era: new maritime routes and geopolitical tensions. In Delfour-Samama, O., Leboeuf, C., and Proutière-Maulion, G. *New Maritime Routes: Origins, Evolution and Prospects*. Paris: A. Pedone.

Dalaklis, D. and Baxevani, E. (2016). Arctic in the global warming phenomenon era: new maritime routes and geopolitical tensions. In Delfour-Samama, O., Leboeuf, C., and Gwenaele Proutière-Maulion, G. *New Maritime Routes: Origins, Evolution and Prospects*. Paris: A. Pedone.

Dalaklis, D. (2017). Safety and security in shipping operations. In Visvikis, I. and Panayides, P. Shipping Operations Management. *WMU Studies in Maritime Affairs*, Vol. 4. Cham: Springer: pp. 197–213.

Dalaklis, D. and Baxevani, E. (2017). Maritime routes in the Arctic: examining the level of traffic and port capabilities along the Northern Sea Route. In Chircop, A., Coffen-Smout, S.; and McConnell, M. *Oceans' Yearbook 31*. Leiden: Brill Nijhoff. pp. 106–135.

Dalaklis, D. and Drewniak, M. (2017a). The Arctic region: mapping the current state of icebreakers and identifying future trends. Maritime Search and Rescue 2017 Conference, Helsinki, Finland.

Dalaklis D. and Drewniak, M. (2017b). The Arctic region: mapping the current state of icebreakers' availability. 7th MARPART Conference: Competence Development within Maritime Preparedness in the High North, Nuuk, Greenland.

Dalaklis, D. and Drewniak, M. (2018). *Trends and challenges in relation to cruising activities in the Arctic*. Maritime Search and Rescue 2018 Conference: Developing Capabilities for SAR Operations, Helsinki, Finland.

Dalaklis, D. and Olcer, A.I. (2018). Polar Water Operations: 'Food for Thought'. *Journal of Ocean Technology*, 13(4), pp. v–vi.

Dalaklis, D., Baxevani, E., and Siousiouras, P. (2018a). The Future of Arctic Shipping Business and the Positive Influence of the International Code for Ships Operating in Polar Waters. *Journal of Ocean Technology*, 13(4), pp. 76–94.

Dalaklis, D.; Drewniak, M. and Schröder-Hinrichs, J-U. (2018b). Shipping operations support in the High North: examining availability of icebreakers along the Northern Sea Route. *WMU Journal of Maritime Affairs*, Vol. 17, No. 2, pp. 129–147.

Dalaklis, D. (2019). Exploring the Issue of Search and Rescue Services in the Arctic, 14th Arctic Shipping Summit, Montreal-Canada, 13 March.

Dalaklis, D. and Drewniak, M. (2019). Examining the Status of the United States' and Canada's Ice-Breaking Fleets, Arctic Coast Guard Seminar 2019, Turku, Finland, 4 April.

Dalaklis, D. and Raneri, P. (2019). Increasing the Level of Safety in the Arctic: Training/Exercise for Key Personnel in Emergency Response, 8th MARPART CONFERENCE 'MARITIME EMERGENCY RESPONSE IN THE ARCTIC' − policies, capacities and competence, 4 September.

Dawson, C. (2014). Arctic cargo shipping volume is rising as ice melts. The *Wall Street Journal*. Retrieved from: www.wsj.com/articles/arctic-cargo-shipping-volume-is-rising-as-ice-melts-1414612143.

Deboer, S. (2017). Arctic Security and Legal Issues in the 21st Century: An Interview with CDR Fahey. Retrieved from: http://cimsec.org/arctic-security-legal-issues-21st-century-interview-cdr-sean-fahey/31016.

DeWitz, J., Dalaklis, D., Olcer, A.I., and Balini, F. (2015). Arctic LNG: Exploring the Benefits of Alternative Fuels to Mitigate Environmental Impact Risks. ShipArc 2015: Safe and Sustainable Shipping in a Changing Arctic Environment (World Maritime University), Malmo, Sweden.

Drewniak, M., Dalaklis, D., Kitada, M., Ölçer, A.I., and Balini, F. (2017). Geopolitical considerations of shipping operations in the Arctic: mapping the current state of icebreakers and identifying future needs. International Association of Maritime Economists 2017 Conference, Kyoto, Japan.

Drewniak, M. and Dalaklis, D. (2018). Expansion of business activities in the Arctic: the issue of search and rescue services. In Chircop, A., Coffen-Smout, S., and McConnell, M. *Oceans' Yearbook 32*. Leiden: Brill Nijhoff: pp. 425–455.

Drewniak, M., Dalaklis, D., Kitada, M., Ölçer, A.I., and Ballini, F. (2018). Geopolitics of Arctic shipping: the state of icebreakers and future needs. *Polar Geography*, Vol. 41, No. 2, pp. 107–125.

Government of Norway. (2017). Hovedredningssentralen Nord - Norge. Retrieved from: www.hovedredningssentralen.no.

International Security Advisory Board. (2016). Report on Arctic Policy. U.S. Department of State (21 September). Retrieved from: https://state.gov/documents/organization/262585.pdf.

Humpert, M. (2014). Arctic shipping: an analysis of the 2013 Northern Sea Route season. The Arctic Institute. Retrieved from: https://arcticyearbook.com/arctic-yearbook/2014/2014-briefing-notes/111-arctic-shipping-an-analysis-of-the-2013-northern-sea-route-season.

Kiiski, T. [2015]. Supply and demand of transit cargo along the Northern Sea Route. In Weintrit, A., *Activities in Navigation: Marine Navigation and Safety of Sea Transportation*. Boca Raton, FL: CRC Press.

Lasserre, F. (2012). Climate change and commercial shipping development in the Arctic (Commercial Shipping). ArcticNet Annual Research Compendium (2011–2012).

Lasserre, F. (2014). Case studies of shipping along Arctic routes. Analysis and profitability perspectives for the container sector. *Transportation Research Part A*, Vol. 66, pp. 144–161.

Lee, T.D. and Kim, H.J. (2015). Barriers of voyaging on the Northern Sea Route: a perspective from shipping companies. *Marine Policy*, Vol. 62, pp. 264–270.

Northern Sea Route Information Office. (2016). Transit Statistics. Retrieved from: http://arctic-lio.com/?cat=27.

Rodrigue, J.P. (2013). *The Geography of Transport Systems*, (3rd ed.). New York: Routledge.

Rodrigue, J.P. (2016). Main routing alternatives between the Pacific and Atlantic. Retrieved from: https://people.hofstra.edu/geotrans/eng/ch1en/appl1en/map_panama_alternatives.htm.

Schøyen, H. and Bråthen, S. (2011). The Northern Sea Route versus the Suez Canal: cases from bulk shipping. *Journal of Transport Geography*, Vol. 19, pp. 977–983.

Seker, B. and Dalaklis, D. (2016). Maritime energy security issues: the case of the Arctic. *Geopolitica Revista*, Vol. XVI, No. 66, pp. 171–186.

5 Activity and Risk versus Preparedness in the High North

Bent-Ove Jamtli

An acceptable level of safety, including a high level of preparedness for undesirable incidents, is crucial for the sustainable development of the Arctic communities and healthy economic development in the High North. All commercial activity in the High North must include a safety and emergency response dimension. Safety is particularly important for operations in the autumn and winter months in the Norwegian Sea, the Barents Sea, and the sea areas around Svalbard. With increased commercial activity related to cruise traffic, oil and gas activities, cargo transport, and changes in the activity patterns related to the fishing fleet, research, and military activity, it becomes important to ensure the further strengthening and continuous presence of public emergency capacities in the region.

The activity in the High North, both on land and at sea, is strongly increasing in summer and winter alike. About 30 new expedition cruise vessels are currently being built and will be put into service in the High North over the next few years. This new equipment will further enhance the already steady rise in cruise tourism in the High North and could therefore inexorably increase the risk of accidents with many lives at stake. In this area with vast distances and few emergency resources, a significant accident at sea would most likely cause severe damage to life and the environment. In addition, land-based summit tours and expedition tourism in the High North are increasing. A fatal accident involving nine Russian tourists who drove snowmobiles and went through the ice on the Tempelfjord in 2017, where one them lost their life, is just one of several examples that show that increased activity raises the risk of accidents.

On top of this fact, climate change in the High North is also noticeable. Higher temperatures and more precipitation in the form of rain lead the permafrost to melt and the soil and rock masses to become unstable. This also builds up larger snow floats in the mountains in the winter than before. In recent years, we have witnessed more land and snow avalanches than usual. In some of these avalanches, the tragic loss of human life resulted.

As a result of less and less sea ice, fishing is now taking place north of Svalbard throughout the year. Less sea ice in the Arctic opens up the space for considerably more maritime activity than before. Unfortunately, increased maritime activity in more remote areas during the whole year, combined with extreme weather conditions, could lead to more fatal accidents, as the consequences of an accident will be much more severe when it happens in a remote place far from the nearest search and rescue (SAR) capacities.

After the larger accidents that have taken place in Svalbard in recent years, emrgency response equipment and crew have had to be collected from the Norwegian mainland. If the necessary operational capacities are large and heavy, it can take several days to transport them to Svalbard. These situations can lead to very long response times, resulting in an increased risk to life and the environment.

What will help to strengthen civilian preparedness in the High North?

The establishment of broadband satellite communications will undoubtedly improve the ability to communicate more precisely and efficiently during emergencies. This implementation is under way and will hopefully be completed by the end of 2022.

The newly announced improvement of high-frequency radio communications in the northern Norwegian Sea, around and north of Svalbard, is also imperative. There are large areas without maritime emergency radio coverage in the High North the moment.

The pre-deployment of emergency equipment along the coast around Svalbard, as well as expanding the emergency preparedness base in Longyearbyen to be able to pre-store emergency drop kits and heavier equipment, will also improve the level of preparedness. This type of equipment takes too long to transport from the mainland in the case of acute accidents.

The Governor of Svalbard will from 2020 have at his disposal a contingency vessel year-round. This vessel should be upgraded to a sufficient level of sea-ice classification and be suitable for the rescue and evacuation of a substantial number of injured and uninjured persons in maritime accidents. There should also be put in place a requirement for a buddy-buddy system for rescue capacity for vessels sailing in the area. Standardized systems for hoisting survivors, lifeboats, and life rafts from the sea should be implemented.

A new rescue helicopter base will be established in Tromsø by the end of 2022. This base will fill the existing gap in maritime emergency preparedness in the north Norwegian Sea and the Barents Sea.

The level of realistic SAR exercise activity in the High North should be increased. Through more exercises, we will reveal weaknesses and deficiencies that can be corrected before the next severe accident occurs. A better-trained emergency response organization will provide improved emergency preparedness through more professional and efficient co-operative rescue efforts. The same co-operation applies to oil spills and environmental damage response.

One should also aim to establish a High North Analysis capacity at the Joint Rescue and Co-ordination Centre (JRCC) in North Norway to be able to achieve a more comprehensive and knowledge-based approach to the challenges that we are facing in the present and foreseeable future, including an 'Arctic lessons learned' database from real emergencies and exercises in the High North.

JRCC North Norway is leading an extensive network of practitioners, academics, and industry partners in the Arctic and North Atlantic called the Arctic Security and Emergency Preparedness Network (ARCSAR). This project is financed with €3.5 million over five years from the EU's Horizon 2020 programme. It has 21 partners from 13 different countries, including all the Arctic circumpolar nations, except Denmark and Greenland. The ARCSAR project will address the Arctic and North Atlantic region as it prepares to cope with the safety threats that will result from increased commercial activity in the region.

JRCC North-Norway is also working closely with the Arctic Council Emergency Prevention, Preparedness and Response Working Group to establish an Arctic SAR rescue co-ordination centre forum for expanded SAR cross-border collaboration to exchange best practices, co-ordinate procedures, reduce language and cultural barriers within the SAR community, and to set up an international database for the seamless exchange of SAR and oil spill response emergency preparedness information between the countries in the Arctic region.

6   Vessels and Opportunities

How the Arctic Expedition Cruise Industry Can Contribute to Enhanced Marine Preparedness in Polar Waters

Frigg Jørgensen and Edda Falk

Ask any mariner, and they will tell you that people are the most precious cargo. In the case of a marine incident, the first and foremost priority will be to safeguard people. This principle is deeply rooted in our systems of marine preparedness and is front and centre when planning operations in polar waters. For the Arctic expedition cruise industry, which carried over 32,000 passengers on Arctic voyages in 2019, ensuring the safety of passengers and crew is paramount. Although these voyages take place during the summer season when conditions are more favourable, the Arctic can be a challenging environment in which to operate at any time. Long distances, limited connectivity, lower temperatures, as well as unpredictable weather and ice conditions, all contribute to making the Arctic a region where you need to come prepared. This situation is especially true when your ship is loaded with people. Expedition-style vessels are smaller than conventional cruise ships, typically carrying somewhere between 150 to 200 passengers. These vessels generally are ice-classed and have crews with many years of experience operating in polar environments. When planning and carrying out these voyages, safety is the number one priority.

Safety and marine preparedness are priorities for the Arctic expedition cruise industry. The vast majority of operators in this segment are part of the Association of Arctic Expedition Cruise Operators (AECO), and many efforts to enhance preparedness are channelled through this organization. AECO is an international organization for expedition cruise operators and associates in the Arctic, dedicated to managing environmentally friendly, safe, and responsible cruise tourism. The Association has 75 international members, including 46 vessel operators, owners and management, and 60 expedition cruise vessels that are organized by the Association. Since its establishment in 2003, AECO has been the forum for dialogue and collaboration between the industry, search and rescue (SAR) entities and relevant authorities. Cross-sector dialogue can contribute to lower risks and a better joint understanding of capabilities and procedures and help operators and SAR responders alike to develop best practices that enable them to work well together in preparing for and responding to marine incidents. However, the goal of this collaboration is not only to lower risks and strengthen preparedness in the expedition cruise sector, but also to focus on how the industry can be an asset in SAR operations and contribute to enhanced marine preparedness in the Arctic. One important forum for this dialogue is the Arctic Security and Emergency Preparedness Network (ARCSAR), which will be discussed later in this chapter.

Expedition Cruise Ships: A Valuable Resource in Arctic SAR

When discussing Arctic cruising and preparedness, the focus can often be on cruise vessels as a potential recipient of assistance. Recent intersectional exercises have shifted focus and highlighted the fact that passenger vessels can represent a very valuable asset in Arctic SAR operations. Expedition cruise ships sail to remote areas of the Arctic during the summer season. They will therefore often be one of the nearest available resources if another vessel or even a local community experiences difficulty and needs assistance. In such cases, the expedition cruise ships can act as a vessel of opportunity to the vessel in distress. As the Arctic is characterized by long distances and limited SAR capabilities, vessels of opportunity can be an invaluable asset in SAR operations in the Arctic. In many cases, a vessel of opportunity will be able to reach a vessel in distress long before SAR vessels or other assistance can arrive on site.

Passenger Ships as Vessels of Opportunity

A vessel of opportunity can be defined as a vessel close enough to aid a vessel in distress. Under international maritime law, every vessel at sea is required to assist in a distress situation. The vessel of opportunity can be released from the obligation to assist if the vessel in distress or SAR responders inform them that their assistance is not needed.

In the summer months, expedition cruise ships sailing in Arctic waters represent a great resource for preparedness and response. Not only are these vessels present in remote locations far from other SAR assets, but the ships also carry equipment, supplies, and personnel that enable them to make a significant contribution to SAR operations. Cruise ships carry food, water, medical supplies, and numerous high-speed small vessels and other resources on board that are useful in SAR operations. Personnel on board expedition cruise vessels may include many trained experts, such as medics, nurses, firefighters, divers, welders, and others. The first step is having these resources in place on a vessel that is in the proximity of the vessel in distress. The second, but equally important step is making sure that these resources are used in the best and most effective way possible. AECO is working with SAR entities and researchers to ensure that all actors have the knowledge, routines, best practices, and contacts necessary to make this happen.

Enhancing Dialogue to Ensure Optimal Use of Available Resources

Tabletop exercises organized by AECO and SAR entities have shown that there is a potential for making better use of the resource that cruise ships represent. Increased dialogue between SAR entities and operators and vessel-owners will give a better understanding of available resources, needs, and operational requirements. AECO is working with Arctic SAR entities to facilitate this dialogue. One important platform for collaboration is the Joint Arctic SAR Tabletop Exercise and Workshop, which AECO initiated in 2016 and which has become an annual event. The event has become part of the Arctic Security and Emergency Preparedness Network (ARCSAR) project. It is organized in collaboration with the Icelandic Coast Guard and the Joint Rescue Co-ordination Centre North-Norway (JRCC North-Norway). The event gathers cruise operators, SAR responders, researchers, and other stakeholders. The objective of the gathering is to strengthen dialogue, co-operation, and the exchange of information between the Arctic cruise industry and various Arctic SAR responders. The events have attracted up to 80 participants representing a high number of AECO members and SAR entities from Canada, the USA, Iceland, Greenland, Denmark, Finland, The Faroe Islands, Svalbard, mainland Norway, and Sweden.

The Joint Arctic SAR event has proved especially useful for commanding bridge officers, cruise operation management, and SAR co-ordinators. Commanding officers and management get the chance to engage directly with the people who will be on the other end of the line if an incident occurs. Simultaneously, the SAR entities get a larger understanding of cruise operations. There have been several instances that can only be characterized as 'Aha!' moments, where the interlocutors from the different sectors realize that they lacked an essential piece of information or had made assumptions that proved to be false. One example of this is SAR responders pondering over the logistics of refuelling a helicopter assisting a vessel in a remote location. Hearing their discussions, the expedition cruise operator in charge of an on-site vessel of opportunity informed them that they had abundant helicopter fuel available on the cruise vessel. By learning that refuelling on site was an option, the SAR responders were saved a detour of several hours.

There were other examples, such as participants being surprised to learn that rescue helicopters can assist with polar bear watch using thermal cameras. Others were not aware that some cruise operators routinely bring tents, emergency stoves, survival kits including medicines, and a ship doctor when they go ashore for excursions. Information such as this will be essential in the case of a real incident.

Experiences from the Joint Arctic SAR Tabletop Exercise and Workshop indicate that there is a lack of knowledge among responders of the capabilities, resources, and equipment that are available on expedition cruise ships. Cruise operators also need to gain a better understanding of the standing operating procedures of SAR responders and what is expected of them if and when they take on the role of vessel of opportunity or on-scene co-ordinator. AECO is working with relevant stakeholders to ensure that those gaps are bridged.

Helirescue Training Video

One example of a successful initiative to prepare cruise operators for assisting rescuers in the case of an incident is the recently launched Helirescue training video. The project is the result of a collaboration between Lufttransport AS – a leading provider of SAR emergency preparedness and services in the Arctic, AECO, and the Governor of Svalbard, as part of the ARCSAR project.

The video explains how the crew and staff on board vessels can assist in a helicopter rescue operation. The video contains instructions on how the person on deck should prepare the hoisting area, receive the guideline that is lowered from the rescue helicopter, follow the instructions of the rescuer, and assist the rescuer in hoisting people from the deck. Cruise operators can use the video for training purposes. These vessels have capable and highly trained staff who can assist rescuers in the field. This video can help to educate and train ship crew and staff and other interested parties.

ARCSAR and Networking with the SAR Sector

In addition to Joint Arctic SAR Tabletop Exercise and Workshop, AECO works with the Arctic SAR sector in several ways. One important initiative is ARCSAR. AECO is one of 21 international partners in the EU-funded network, which is led by the JRCC North-Norway. The network brings together SAR and oil spill response entities and industry partners from different sectors: cruise, shipping, and offshore and satellite technology, in addition to research scientists working across sectors to strengthen co-operation and innovation in security and emergency response in the Arctic and the North Atlantic. The ARCSAR project will run for five years and include a live exercise on a cruise vessel. Among other things, the ARCSAR project has launched an online Innovation Arena to identify challenges within Arctic SAR and collect ideas for potential solutions.

AECO's SAR efforts also include close dialogue with the US and Canadian Coast Guards, the Arctic Coast Guard Forum, and the Arctic Council Emergency Prevention, Preparedness and Response Working Group.

Other Measures to Enhance Marine Safety and Preparedness

As outlined above, AECO has several collaborations focused on enhancing preparedness in case of incidents. In addition to these efforts, AECO is working broadly to prevent incidents from happening in the first place. Through several initiatives, AECO is working to enhance safety by focusing on areas such as charting, piloting, crowdsourcing of hydrographic data, the collection of historical sailing data, live vessel-tracking, information flow, the co-ordination between vessels sailing in Arctic waters, and ice and meteorological services.

AECO represents expedition cruise operators with decades of experience navigating in Arctic waters. The association draws on the knowledge, experience, and commitment of its members to establish best practices and high operational standards. AECO's members agree that expedition cruises and tourism must be carried out with the utmost consideration for the Arctic's fragile, natural environment, local cultures, and cultural assets, as well as the challenging safety hazards at sea and on land.

AECO is both a driving force and an industry contact point when it comes to the development of official charts and pilotage service in Svalbard. The association is an observer to the International Hydrographic Organization and engages directly with relevant authorities to convey the needs and preferences of operators sailing in the Arctic.

AECO co-ordinates a crowdsourced collection of depth soundings in Arctic water. Cruise vessels contribute by collecting depth soundings during their voyages. The collected data is shared with operators and authorities and serves as a supplement to official charts.

AECO members must submit their sailing plans well ahead of the start of the season. A booking system for landing sites ensures that overcrowding is avoided and wilderness etiquette is respected. However, it also has the benefit of giving an overview of where AECO members are operating. In addition to sharing their planned itineraries with all other members, members of AECO operating vessels carrying more than 12 passengers are obliged to share their position and trajectory when sailing in the Arctic using a live vessel-tracker. An online portal gives all members access to real-time information about the position of other vessels, which is valuable for voyage planning and safety risk assessments. At the beginning of the season, AECO distributes an updated contact list with contact information for all vessels, so that the ships can contact each other in the field. AECO also shares this information with relevant authorities that request access.

AECO maintains an online database with updated information and historical records on technical specifications of vessels, cruise itineraries, completed voyages and post-visit reports. The database allows members to co-ordinate their sailing plans, and it is also used to generate statistics that are useful for stakeholders such as industry, policymakers, and SAR entities.

As the primary representative of the Arctic expedition cruise industry, AECO liaises with national authorities to assert the industry's need for nautical charts, ice charts, pilotage services, and pilot certification. AECO has been a driving force in the expansion of pilotage service to new areas in Svalbard, and in ensuring real-time ice information and ice-metocean outlooks offered by meteorological institutions.

AECO's members have committed to measures aimed at preventing adverse environmental impacts in the case of an accident. AECO has banned the use and carriage of heavy fuel oil (HFO) by AECO members sailing in the Arctic, as this type of fuel is extremely challenging to recover in case of a spill, compared with lighter distillate fuels. While there are discussions to establish an international legal ban on HFO in the Arctic, the Arctic expedition cruise industry has signalled its commitment to responsible practices by implementing this self-imposed ban.

AECO has also developed several guidelines, some of which are tailored to passenger vessels sailing in the Arctic. The guidelines establish the measures that operators should take to ensure safety during operations and landings.

Conclusion

There is a lot to be gained by ensuring good dialogue and collaboration between the private sector and SAR authorities. AECO's long-standing practice of working with SAR entities has demonstrated that this kind of co-operation benefits both industry and responders. There is also value in inviting research institutions and authorities to be part of these exchanges to help to inform the necessary research and informed decision-making. AECO's experience shows that there is a great willingness both among operators and SAR responders to take advantage of forums of dialogue and exchange that are made available to them. AECO has also seen that mariners, cruise executives, and SAR responders use these forums for frank discussions, candid observations, and sharing details of 'near-misses' and lessons learned. Having platforms that allow open sharing in a trust-based environment is essential in helping both the shipping industry and responders to understand the limitations of their current practices and to develop new standards and routines to meet the challenges that are identified. Far from keeping its cards close to its chests and going about its business, the expedition cruise industry is interested in sharing and willing to listen. Having exchanges such as those facilitated by the Joint Arctic SAR Tabletop Exercise and Workshop and the ARCSAR network is in the interest both of the industry and SAR authorities. It all comes back to the shared principle: People are the most precious cargo. By working across sectors, the expedition cruise industry and the SAR sector can make sure that we are doing all we can to keep people safe when sailing in Arctic waters.

7 Legal Possibilities to Police the Arctic Ocean

Tor-Geir Myhrer

Introduction

This chapter looks at the legal possibilities and professional requirements for policing the Arctic Ocean in order to prevent terrorism and other serious crimes[1]. It deals with how international law makes it possible to prevent or punish acts that are or might be a security threat. Only the areas closest to Norway – the Barents Sea, the Greenland Sea and the Denmark Strait – are covered, but hopefully, this information can have a more general application.

Transnational crimes are not a new issue (Roth, 2014). What has changed is the scale of these activities caused by the deregulation of commerce and breakthroughs in technology, communication, and transportation. In this context, it is a challenge that 70% of the Earth is water. The sea is open, to a large extent, to everyone, with no general or legitimate power to police it. Criminals can misuse this freedom of navigation on the high seas in two different ways: Using the high seas to commit crimes like transporting prohibited substances or irregular migrants. The latter has been illustrated by desperate families trying to flee the horrors of civil war in Syria and difficult living conditions in several countries in North Africa. The second way of misusing the sea is criminal attacks on the peaceful and legitimate use of the ocean as a highway for, as well as the exploitation of the oceans' living resources and seabed. The hijacking of ships, taking hostages, terrorist attacks on ships, and oil/gas-producing platforms are examples of this. In the past two decades piracy has re-emerged as a 'prominent' crime at sea, recently observed on the coast of Somalia and in the Bay of Aden.

Regarding navigation and the climate, there are many differences between the Mediterranean Sea and the coast of Somalia, on the one hand, and the Arctic Ocean on the other. However, the Intergovernmental Panel on Climate Change report from 2014 holds it likely that we will have a nearly ice-free Arctic Ocean in September by about 2050. Combined with technological progress like enforced ships and platforms, stronger icebreakers, and better navigation and communication equipment, this will probably increase both Arctic travel and exploiting of resources in the Arctic Ocean (Marchenko et al., 2015)[2]. Nevertheless, it will also make the area more exposed to (transnational) crimes.

In an ocean where policing until now has first and foremost has been concerned with activity related to the exploitation of living resources such as fishing, searching for oil and gas, or environmental crimes, could soon face new challenges: A sea route for the smuggling of migrants, narcotics, and weapons from the East, and possibly also a maritime area for carrying out terrorist attacks on cruise ships or platforms for oil/gas production.

In the aftermath of the terrorist attacks on New York and Washington in September 2001, many people were optimistic about the ability of the international community to agree on further encroachments on the sovereignty of both flag states and coastal states, to make way for more effective policing of the seas. (Jesus, 2003; Roach, 2004) However, the result of the amendment of the SUA Convention (The Convention for the Suppression of Unlawful Acts and Violence against the Safety of Maritime Navigation) in 2005 showed that this optimism was largely unfounded.[3] Despite the setback in 2005, there is an improved willingness among states to co-operate in developing new rules (see Klein (2011). However, it is still a complicated and time-consuming process to change the Law of the Sea regarding the right to shipping interdiction on the high seas. Moreover, it is the general Law of the Sea that must be amended. Unlike the Antarctic, no special treaties regulate activity in the Arctic region. The Antarctic Treaty of 1959 and other international regulations for the area were necessary because the Antarctic is a continent owned by no country, but where many nation states have active interests. The Arctic region is different. The North Pole is not a continent, but an ocean surrounded by sovereign states. It can be compared to the Mediterranean Sea, disregarding the climatic differences. However, even if the struggle for new rules should be successful, building the capacity and the competence to carry out valid policing of the Arctic Ocean will take time. Taking this into consideration, it is probably not premature to reflect on how international law makes it possible to police the Arctic Ocean. The ice melting could turn out to be faster than the law amendment process.

Theory and Methodology

With the exception of environmental crimes and illegal, unreported, and unregulated (IUU) fishing, the need for policing and crime-fighting in the Arctic Ocean is, at present, a non-existing problem. This article therefore contains no empirical research, but is based on a hypothetical scenario. Unlike the Antarctic, there is no international, regional treaty relevant for policing the Arctic Ocean. The questions arising from the hypothetical scenario will accordingly only be discussed concerning the following multinational conventions: the United Nations Convention on the Law of the Sea (UNCLOS) of 1982, the Convention for the Suppression of Unlawful Acts of Violence against the Safety of Maritime Navigation (SUA Convention) of 1988 with 2005 Protocol, the United Nations Convention against Illicit Traffic in Narcotic Drugs and Psychotropic Substances of 1988, and the United Nations Convention against Transnational Organized Crime of 2004 and the

Protocols thereto. Regional and bilateral agreements will not be discussed. Such conventions will only be relevant if both the coastal state and the flag state are party to the convention. Moreover, if the Arctic Ocean becomes ice-free, the flag states could be every nation in the world. Having my work experience from the police and public prosecutions service, and my academic production mainly on Norwegian criminal law, criminal procedural law and police law, the discussion is based on how leading academics in the field have understood and outlined these conventions. First and foremost, Robin Churchill, A.V. Lowe, Douglas Guilfoyle and Natalie Klein.

Delimitations and Definitions

Which part of the Arctic Ocean?

The Arctic Ocean is located within the Arctic region. However, three different criteria are used to define the region: The area north of the Arctic (Polar) Circle; the area north of the 'tree line'; and the area north of the isotherm, where the average temperature is lower than 10°C during the warmest month (July). In this article, the isotherm is used as the criterion.

This article focuses on the waters closest to Norway. It covers the area from the southern point of Greenland (around 40°W) to Novaya Zemlya (around 50°E). This geography includes first and foremost the Denmark Strait, the Greenland Sea, and the Barents Sea. The Northwest Passage (NWP) through the Davis Strait, Baffin Bay, the disputed area of Beaufort Sea (Byers and Baker, 2013), and the Northern Sea Route (NSR) across the Kara Sea and the Laptev Sea (both routes ending in the Bering Strait and the Bering Sea) are not included. It could be argued that with this limitation, the most interesting areas of the Arctic Ocean are the North East Atlantic Fisheries Commission (NEAFC).[4] Moreover, even if important areas of the Arctic Ocean are left out, traffic through the two sea routes will to a large extent still be passing the part of the Arctic Ocean covered in this chapter – more so for the NSR than the NWP.

The legal regulations regarding the right to police the ocean are to a large extent a compromise between the flag states' interests in freedom of navigation, and the coastal states' legitimate interest in knowing and having the right to decide on the activity and the exploiting of resources outside their coast. The coastal states in this chapter are Greenland/Denmark, Iceland, Russia, and Norway.[5] Taking into consideration that the Arctic Ocean in the future could be ice-free, the flag states could be every nation in the world with a fleet, whether coastal or land-locked.

Which Crimes?

The purpose of this article is to discuss to what extent international law makes it possible to discover, prevent, or punish acts that are or might be a security threat. As discussed by Natalie Klein (2011), there is no consensus on which crimes should be considered as a threat to maritime security. Whether IUU-fishing and environmental crimes should be regarded as a threat to maritime security seems to be most disputed. These crimes are not included in this paper. The right to interdict and visit exclusive economic zones (EEZ) and fishery zones are discussed very briefly, even though the legal framework does not resemble multinational treaties, which are the topic of this chapter. Using the right to inspect without suspicion in the fishery zones in order to 'accidentally' find evidence of other crimes besides illegal fishing, raises some ethical questions, although these are not discussed in this chapter. What will be examined here are the possibilities of policing the security threats stemming from piracy or armed robbery at sea, hijacking and taking hostages, sabotage, and the capture of vessels and platforms, human trafficking/smuggling, drugs-trafficking, and weapons-trafficking, including weapons of mass destruction, and terrorism.

Which Maritime Zones and which Policing Activity?

Policing of all maritime zones outside the internal waters will be discussed, including the territorial sea, the contiguous zone, the EEZ, the high seas, and including the special fishery zones around Jan Mayen and Svalbard. The chapter will discuss the right to surveillance, demanding identification, visit and boarding, carrying out searches and seizures, and arresting vessels and crew.[6] In making a framework for the discussion, both types of crimes, maritime zone and policing activity could be used as leading criteria. The core of the policing activity is 'the maintenance of order and security, through the prevention and detection of crime and incivilities' (Mawby, 2007). As the question is to what extent the Law of the Sea makes policing possible, i.e. prevention and detection, I have chosen to discuss policing activity.

A Hypothetical Scenario

By 2050 intelligence strongly indicates that a terrorist attack will be carried out in the eastern part of the Arctic Ocean during late summer. The target is either a cruise ship voyaging around Svalbard or oil-producing installations in the Barents Sea. The attack is likely to be carried out by persons sympathizing with a minority in an Asian republic on the brink of civil war, where the minority feel forgotten and let down by the international community. The attack will be financed by using the ice-free NSR to smuggle drugs (see Roach, 2004) and/or migrants to the northern part of Europe. Neither the identity of persons involved nor vessels to be used are known.

If the Law of the Sea remains unchanged, what are the legal possibilities for the law enforcement authorities in the area to discover, identify and visit a suspected vessel, and eventually carry out search and seizure, and in the end, arrest the crew and vessel?

Surveillance and Information-gathering

To prevent crimes on land, fundamental police methods are patrolling, watching, and gathering information on persons and activities. These activities are also important at sea. The coastal states can do what they like with boats, helicopters, or airplanes within their own territorial waters, as long as it does 'not hamper the innocent passage of foreign ships' (See UNCLOS articles 19 and 24 which defines what kind of activities deprives the status of innocent passage, and which forbids the costal state to discriminate ships of any state or impose requirements on ships which have the effect of denying or impairing the right of innocent passage).

Looking at the question from the opposite angle might be more interesting: Can a coastal state use the right to the innocent passage to carry out surveillance in other coastal states' territorial sea?[7] If people are travelling on foot, it is relatively uncomplicated to control who is crossing the border and enter the country on land. The Russian/Norwegian land border is an example of that. At sea, it is more complicated, and a coastal state could wish to know who is approaching their territorial sea border. So if the NSR in the future will be ice-free, the question is whether the Norwegian Coast Gard can use the right to innocent passage to enter Russian territorial waters to observe which ships are approaching Norwegian waters from the Pacific Ocean. Given the present Convention on the Law of Sea, the answer is clearly 'no'. Even though the vessel in one interpretation of the word could be seen as passing through Russian territorial water when entering it outside Kirkenes and leaving it after passing the Kola Peninsula, it will probably not be regarded as a passage, according to UNCLOS article 18, paragraph 2. This article demands that the 'passage shall be continuous and expeditious'. Furthermore, even if it should be seen as 'passage' in this context, it will most certainly not be seen as 'innocent'. The purpose of surveillance and gathering of information will probably be seen as an 'activity not having a direct bearing on the passage' (see UNCLOS article 19, paragraph 2 [l]) and as such 'be considered to be prejudicial to the peace, good order or security of the coastal State' (Klein, (2011). The conclusion seems to be based on the following premises:

'For intelligence gathering in the territorial sea or straits, interpretations of permissible activities should augur in favour of the exclusive rights of the coastal state consistent with the sovereignty of states in these maritime areas, and in view of the fact that intelligence gathering is most likely for the national defence of the third state rather than responding to collective maritime security threats' (Klein, 2011).

This information could be looked upon differently in the future if the purpose of such surveillance is the prevention of crimes that will enhance the collective maritime security. However, the experiences with the SUA protocol in 2005 do not give cause for much optimism. However, in addition to this problem, there might also be another obstacle. In Norway, as in many other countries, the coast guard is a part of the navy, and as such the ships are 'warships' according to the definition in UNCLOS article 29. Even if the general view is that the right to innocent passage also includes warships, many states demand prior authorization or notification for such passages, see (Churchill and Lowe, 1999).

Like air space, the oceans are impossible to control by patrolling and making observations alone. Airplanes have a mandatory identification and tracking system. An airplane approaching a state's territory without this system or otherwise not identifying itself will be denied entry, forced away by fighters from the air force, or ultimately shot down. At sea, it is different, even though significant improvements have been made in recent years.

In 2006 The International Convention for the Safety of Life at Sea (SOLAS) in chapter V –Safety of navigation – acquired a new regulation, 19–1, on Long Range Identification and Tracking of Ships (LRIT); it entered into force in 2009. According to points 2.1 and 4.1 of the regulations, they apply to the following types of ships on international voyages: Passenger ships, cargo ships of more than 300 gross tonnage and mobile offshore drilling units, but only if the ship was constructed after 31 December 2008 or, if constructed before, is certified for operations in defined sea areas.[8] Such ships shall automatically transmit the identity and position of the ship and the date and time of the position provided (point 5). The coastal states are entitled to receive this information when the vessel is navigating within a distance not exceeding 1,000 nautical miles off the coast unless the ship is located within the territorial sea of another state (point 8.1.3.). But even though the flag state is a signatory to SOLAS, and the ship is of the relevant type, the administration of the flag state is, according to point 9.1, entitled to '… in order to meet security or other concerns, at any time, to decide that long-range identification and tracking information about ships entitled to fly its flag shall not be provided …'.

One might think that if such a decision is made, the consequence could be that the coastal state denied the ship access to their territorial waters. Nevertheless, that is not the case. From point 1 of the regulations it follows that nothing in the regulation: '… shall prejudice the rights, jurisdiction or obligations of States under international law, in particular, the legal regimes of the high seas, the exclusive economic zone, the contiguous zone, the territorial sea or the straits used for international navigation and archipelagic sea lanes'.

One could say that LRIT regulation is a regulation for the 'good guys', and it is easy to agree with the following quotation from Klein (2011):

'The implication here would seem to be that if a vessel is not providing information to a coastal state by a decision of the flag state of that vessel, no action should be taken by the coastal state to treat that vessel as acting suspiciously and interfere with its passage'.

Outside the territorial seas, the right to surveillance and information-gathering creates few legal questions. The freedom of the high seas comprises both coastal and land-locked states' freedom of navigation and overflight (see UNCLOS, article 87). Moreover, according to UNCLOS article 58, paragraph 1, although with some limitations, the same applies to the EEZ. There are some disputes about what kind of restrictions, if any, that apply to military activity in other states' EEZ (see Churchill and Lowe, 1999; Klein, 2011).

However, the right of the coast guard to patrol the area with vessels defined as warships has never been questioned.

Demanding Identity (identification)

Can the coastal state as a condition for passing the territorial sea demand that nationality and registration data are either visible or available upon request? The question cannot be answered by a simple 'yes' or 'no'.

The first reason for this is that the coastal state's legislative power in the territorial sea is disputed and unclear (see Churchill and Lowe, 1999). For vessels on innocent passage[9] the safest position seems to be that the coastal states only can impose legislation for the purposes listed in UNCLOS article 21, paragraph 1, most importantly for navigational, fishery, and pollution reasons. Security reasons are not mentioned, but according to UNCLOS article 19, a passage is not innocent if it is 'prejudicial to the peace, good order or security of the coastal state'. To see it as a threat to the security that a ship is passing territorial waters without the coastal state knowledge of its full identity is most likely not acceptable. This position is not undisputed. However, in this context, it is also important that many states allow smaller vessels to fly their flag without formal registration if the vessel is owned by a national of the flag state (see Guilfoyle, 2009). At this point, a reminder of earlier discussion concerning flag states decision not to transmit LRIT signals is relevant. As Klein (2011) put it, this did not allow 'the coastal state to treat that vessel as acting in a suspicious manner and interfere with its passage'.

However, the coastal state must always be allowed to demand that the vessel reveals its nationality when in the innocent passage, i.e. showing its flag. As expressed earlier, police interventions at sea are largely based on a balance between the rights of the coastal state and that of the flag state. If there is no flag state, there is nothing to balance. Moreover, it follows from UNCLOS article 110, paragraph 1e that even on the high seas, every warship that encounters a foreign ship refusing to show its flag can board the ship. In territorial waters it must follow that a passage can be considered as not innocent if the vessel refuses to show its flag, as the passage is not taking place 'in conformity with this Convention [UNCLOS] and with other rules of international law', to use the wording of UNCLOS art. 19, paragraph 1.

The legal position might be summarized like this: The coastal state cannot always demand the full identity of vessels passing their territorial waters, but it can always demand to know which nation who is responsible for it.

The Right to Interdict and Visit (board or inspect)

Delimitations and introduction

The right of the coastal states to board and subsequently carry out searches and seizures on board a vessel is, as already mentioned, based on a balance between the interest of the coastal state and the right to freedom of navigation for the flag state. It follows that if it is reasonable to suspect that the vessel is without nationality, the coastal state will have a more unrestricted right to board. (see UNCLOS article 110, paragraph 1d). However, the discussion below will concentrate on how the rights of the flag state and coastal state are balanced.

As mentioned above, the discussion will be based on multinational conventions of which most maritime nations are part. The United Nations Convention on the Law of the Sea (UNCLOS), the United Nations Convention against Transnational Organized Crime and the Protocols thereto (UNTOC), the United Nations Convention against Illicit Traffic in Narcotic Drugs and Psychotropic Substances, the International Convention for the Safety of Life at Sea (SOLAS), and the Convention for the Suppression of unlawful Acts of Violence against the safety of Maritime Navigation with Protocol (SUA Convention). Some regional agreements, like the Council of Europe Agreement on Illicit Traffic by Sea and The Northwest Atlantic Fisheries Organizations Conservation and Enforcements Measures, will also be discussed.

Whether a vessel can be boarded legally depends mainly on two decisive factors: What kinds of crimes are suspected, and in which maritime zone the boarding shall take place. The leading factor in the presentation below will be 'what kind of crime'.

Piracy

UNCLOS article 110, paragraph 1 allocates a general right to board a foreign ship when there are reasonable grounds to suspect three types of crimes: Piracy, the slave trade, and unauthorized broadcasting.[10,11] Because of this general right to board and subsequently carry out further examinations on board, compared with the more limited rights following from the SUA Convention when terrorist crimes are suspected, an interpretation where as many as possible grave violent crimes at sea fall within the definition of piracy is desired.

> According to UNCLOS article 103, letter a) piracy consists of: ...any illegal acts of violence or detention, or any act of depredation, committed for private ends by the crew or the passengers of a private ship or a private aircraft and directed:
>
> (i) on the high seas, against another ship or aircraft, or persons or property on board such ship or aircraft;
> (ii) against a ship, aircraft, persons or property in a place outside the jurisdiction of any State.

A detailed interpretation of this definition is not called for here, but some reflections on three of the elements are appropriate.

First, piracy must involve two ships. If our hypothetical terrorist does not plan to attack the cruise ship from their own vessel but enter the ship as passengers to launch a suicide attack, the intervention cannot be based on UNCLOS article 110. This interaction is an *Achille Lauro*[12] scenario – the situation that gave rise to the SUA Convention.

Second, piracy in the framework of UNCLOS cannot take place in territorial waters. An attack here could be 'armed robbery at sea'. However, this passage will obviously not be innocent, and the coastal state will have full jurisdiction to board, search, and arrest. The wording in the definition is 'a place outside the jurisdiction of any State'. The coastal states will have some jurisdiction both in the contiguous zone and in the EEZ. Does this also rule out piracy as understood in UNCLOS? Based on the preparatory work to UNCLOS article 103 and the wording of article 58, paragraph 2, the answer to this question is 'no', (see Guilfoyle, 2009.

The third element poses the most controversial question: What is the meaning of 'committed for private ends'. The question is whether 'private' shall be understood as opposed to 'public', or as opposed to both 'political' and 'public'. The first interpretation excludes only attacks which have a public sanction from the state of the perpetrators, but the latter will also exclude politically motivated attacks. Douglas Guilfoyle (2009, 2013, 2014) has strongly argued for the first interpretation (see also Frostad, 2016). For the time being, the safest position seems to be that piracy in the meaning of UNCLOS also excludes attacks with political motivation. See Klein (2011) and Jesus (2003: 377–79). If so, the consequence will be as follows: Even if our hypothetical terrorists plans to attack the cruise liner in international waters on its way to Svalbard, their vessel cannot be boarded and intercepted as pirates with legal authority in UNCLOS article 110, as in the definition of the convention they are not pirates, as their motivation is political.

The Slave Trade, Trafficking in Persons and Smuggling of Illegal Immigrants

As already mentioned, UNCLOS Art. 110 authorises a general right to board vessels engaged in the slave trade. However, unlike the situation regarding piracy, the convention has no definition of "slave trade". It seems generally accepted that "slave trade" only encompasses situations where ownership and buying and selling of people are involved. Even if situations where slave trade has been observed in modern times, ((Guilfoyle, 2009: 75–77), this is considered a 'historical crime'.

The UNCLOS slave trade regulations are not applicable for what is called the modern form of slavery – trafficking in persons – where victims are recruited and transported by the use of threats, force, and other forms of coercion, abduction or deception. The appropriate international legal instrument in these situations is the Protocol to Prevent, Suppress and Punish Trafficking in Persons, Especially Women and Children, supplementing the United Nations Convention against Transnational Organized Crime.

However, this protocol contains no regulation of the possible right for a coastal state to board a foreign vessel. The reason for this seems to be that trafficking of people across international waters seldom takes place. Traffickers normally use land or air transport, and the victims cross the border with genuine (or false) travelling documents. So if an ice-free NSR is to be used for trafficking people from Asia to Europe, thus entering the continent in Tromsø, it will be a rather complicated and time-consuming task to obtain the flag state's authorization to board the ship before it enters the contiguous zone; at which point the only option left will often be to escort the vessel to its intended destination.

If there are reasonable grounds to suspect that the vessel is engaged in the smuggling of migrants only – meaning for financial benefit to procure persons to cross the borders without complying with the necessary requirement for legal entry into the receiving state – the legal situation is surprisingly slightly better. The Protocol against the Smuggling of Migrants by Land, Sea and Air, Supplementing the United Nations Convention against Transnational Organized Crime, states in article 8, paragraph 2a the following:

> A State Party that has reasonable grounds to suspect that a vessel exercising freedom of navigation in accordance with international law and flying the flag or displaying the marks of registry of another State Party is engaged in the smuggling of migrants by sea may so notify the flag State, request confirmation of registry and, if confirmed, request authorization from the flag State to take appropriate measures concerning that vessel. The flag State may authorize the requesting State, inter alia:
>
> To board the vessel;
>
> ...

It requires no further explanation that the possibility to obtain an authorization from the flag state is far from certain. Even if the authorization is granted, it will unquestionably be time-consuming: First, waiting for confirmation of registry and then for authorization. That the time differences between the requesting state and the flag state often have to be taken into consideration does not make it any better. When the smuggling of migrants is suspected, then interdiction, when the vessel enters the contiguous zone, could be the earliest legal opportunity. At this point, it will often be more convenient to let the vessel proceed to the destined harbour and take action there.

Terrorism

The abstract to José Luis Jesus's article 'Protection of Foreign Ships against Piracy and Terrorism at Sea: Legal Aspects' (Jesus, 2003) begins: 'This presentation gives a factual assessment of the phenomenon of piracy and other forms of terrorism at sea ...'

Even if there is no generally accepted definition of 'terrorism', it could be pertinent to see most pirates as terrorists. However, as shown earlier in the

'Piracy' section, not all terrorists at sea are pirates. For many unlawful and violent acts against ships at sea, legal authority for interdiction has to be found in the Convention for the suppression of unlawful acts of violence against the safety of maritime navigation (SUA Convention) from 1988, as amended and supplemented by the Protocol of 2005.

SUA Convention article 3, as amended and supplemented by article 4 in the Protocol of 2005, makes it an obligation for state parties to criminalize an extensive list of actions and to establish jurisdiction over these offences. The most significant offences covered in the conventions are violence against or destruction of ships, violence against persons on board endangering the safe navigation of the ship, the seizure or exercising control over a ship by force or threat thereof, and placing or causing to be placed any device or substance that is likely to destroy or cause such damage to the ship or cargo that it endangers its safe navigation. Furthermore, attempts, participation, or organizing such crimes is an offence and must be criminalized.

However, the important question at this point is what right a coastal state has to board a vessel when there are reasonable grounds to suspect that a crime covered by the SUA Conventions is about to be committed? Jesus (2003) argued in favour of the assertion that terrorist attacks at high sea should be treated the same as piracy:

> If, in the case of piracy, the jurisdiction of any state to police foreign ships on the high seas (and therefore to board, search, seize and arrest offenders and ship), has long been accepted and if its operation for centuries has not brought any significant damage to the prerogatives of the flag state, I see no valid reason why it should not be accepted in the case of maritime terrorism.

However, as Klein (2011) puts it:

> Ultimately, however, the 2005 SUA Protocol creates a much tighter balance in that the protections for the flag state do not include all those mentioned by Jesus but nor does it establish by dint of the treaty alone a new exception to exclusive flag state jurisdiction.

It follows from SUA Convention article 8, paragraph 5 that whenever a state party encounters a ship flying a flag of another state party and has reasonable grounds to suspect that the ship is involved in the commission of an offence covered by the convention, and for that reason wish to board the vessel, it shall request (...) that the first party confirm the claim of nationality. If nationality is confirmed, the requesting party shall ask the first party (hereinafter referred to as 'the flag state') for authorization to board.

The primary procedure is thus to require an often rather time-consuming, ad hoc authorization from the flag state. The procedure is quite similar to that of smuggling of illegal immigrants (see above), even though a terrorist attack at sea is undoubtedly a more serious crime. When requested, the flag state has three options: It can authorize the request, or it can conduct the boarding itself (alone or with the requesting state). However, even though the flag state, according to article 13 of the Convention, must co-operate in the prevention of offences outlined in the Convention, it can also decline to authorize the request for boarding. If they do so, article 8bis, paragraph 5c-iv of the Convention explicitly states that: 'The requesting Party shall not board the ship or take measures set out in subparagraph (b) without the express authorization of the flag State'.

The Convention, however, also allows for two other procedures to achieve authorization to board a foreign ship. According to article 8bis, paragraph 5d, a state party can notify the Secretary-General of the International Maritime Organization that authorization should be seen as granted 'if there is no response [...] within four hours of acknowledgement of receipt of a request to confirm nationality'. In addition, according to article 8bis, paragraph 5e, flag states party to the convention can also in advance grant comprehensive permission to other parties to board. Such a notification can be withdrawn at any time.

Unless our hypothetical terrorists use a vessel from a flag state that has accepted one of the two additional procedures, getting authorization to board could be both an uncertain and lengthy mission.

Drug-trafficking

Even though UNCLOS in article 108 recognizes illegal trafficking of narcotic drugs as a major concern and instructs all states to '...cooperate in the suppression of illicit traffic in narcotic drugs and psychotropic substances engaged in by ships on the high seas contrary to international conventions', the coastal states are in article 110, like piracy and the slave trade, not granted any right to visit a vessel suspected of engaging in such illicit traffic.

Such a ruling is given in article 17 in the United Nations Convention against Illicit Traffic in Narcotic Drugs and Psychotropic Substances of 1988. What is granted here is not an actual right to visit like that in UNCLOS article 110, but a procedure for the flag state following a request to authorize the coastal state to interdict and board. A regulation we recognize from both the UNTOCs protocol on the smuggling of illegal immigrants and the SUA Convention on terrorism. UN Drug Convention article 17, paragraph 3 states:

> Party which has reasonable grounds to suspect that a vessel exercising freedom of navigation in accordance with international law, and flying the flag or displaying marks of registry of another Party is engaged in illicit traffic may so notify the flag State, request confirmation of registry and, if confirmed, request authorization from the flag State to take appropriate measures in regard to that vessel.

The first registry must be confirmed, and then authorization to board can be requested. article 17, paragraph 4, deals with the flag state's response to such a request:

In accordance with paragraph 3 (...), the flag State may authorize the requesting State to, *inter alia*:
　Board the vessel

Two points are noteworthy:

Even though article 17, paragraph 1 demands that 'the parties shall co-operate to the fullest extent possible to suppress illicit traffic by sea, in conformity with the international law of the sea', the word 'may' shows that the flag state has no obligation to authorize such a boarding or conduct or participate in it themselves.

When trafficking in drugs is suspected, time is very much of the essence, as evidence easily can be destroyed by dropping the cargo overboard. In article 17, paragraph 7, the flag state shall respond expeditiously, but no time limit is set. Nevertheless, if both the requesting coastal state and the flag state are parties of the Council of Europe Agreement on Illicit Traffic by Sea of 1995[13], it follows from article 7 in the agreement that an answer must be given within four hours after receiving the request. However, not even the Council of Europe countries could agree that no answer should be seen as a tacit authorization (see (Klein, 2011).

The Right to Visit within the Contiguous Zone

What has been discussed above is the right to visit a foreign ship on the high sea, including the EEZ. In the contiguous zone, extending 24 nautical miles from the baseline and accordingly 12 miles outside the territorial sea, the coastal state has a more extensive right to visit. According to UNCLOS article 33, paragraph 1a, '... the coastal State may exercise the control necessary to: prevent infringement of its customs, fiscal, immigration or sanitary laws and regulations within its territory or territorial sea;'

From this follows that the coastal state cannot police traffic in the contiguous zone for security purpose in general, (see Churchill and Lowe, 1999). Certain limitations are relevant in this context. It is generally accepted that the coastal state does not have legislative power in the contiguous zone (see Churchill and Lowe, 1999). It follows from this that the coastal state, according to UNCLOS article 33, paragraph 1a, can prevent crimes only that will affect its territory, including the territorial sea. This effect means typically that the UNCLOS article will not give legal authorization to interdict and visit a vessel that only is passing through the contiguous zone with no intention to enter the coastal state territorial sea.

Not all infringements of the coastal state laws and regulation can legally be prevented by interdicting and visiting foreign ships in the contiguous zone. The 'crime' has to be related to customs, fiscal, immigration, or sanitary laws and regulations. However, in our context, it means that vessels suspected of being engaged in trafficking in drugs, persons, or illegal immigrants can be interdicted and visited without authorization from the flag state. As Klein

(2011) points out, this could also be relevant for combatting terrorism, if the persons engaged in such activities try to smuggle contraband (or merchandise) into the country for later use.

With reasonable grounds for suspicion of such a potential infringement on the coastal state laws, the preventive right to interdict and visit foreign vessels does not seem to be disputed. However, as I will return to later, the right to arrest and escort the vessel to a harbour in the coastal state could be restricted by a proportional clause (see Klein (2011).

The Right to Interdict and Visit in EEZs and Fishery Zones

Even though illegal, unreported or unregulated fishing (IUU-fishing) and other environmental crimes are not discussed in this chapter, the rights of the coastal states in EEZ and fishery zones are not without relevance. A considerable part of the Arctic Ocean covered in this chapter, is covered by EEZ and fishery zones belonging to the coastal states Russia, Norway (including Svalbard and Jan Mayen), Iceland, and Greenland (Denmark). The genuine high sea areas are limited to one area located between the Norwegian EEZ and the fishery zones around Svalbard and Jan Mayen, and the second between the fishery zone around Svalbard and the Russian EEZ outside Novaya Zemlya.[14]

For trawlers and other fishing boats UNCLOS article 73, paragraph 1 stipulates that a coastal state:

> ...may, in the exercise of its sovereign rights to explore, exploit, conserve and manage the living resources in the exclusive economic zone, take such measures, including boarding, inspection, arrest and judicial proceeding, as may be necessary to ensure compliance with the law and regulations adopted by it in conformity with this Convention.

The same must apply for the fishery zones around Svalbard and Jan Mayen. In the genuine high sea areas, a more limited right to board follows from the Convention on Future Multilateral Co-operation on North-East Atlantic Fisheries (NEAFC) of 1980[15] and the NEAFC Scheme of Control and Enforcement of 2014. From Chapter IV in the Control and Enforcement Scheme, it follows that parties to the NEAFC accept a mutual right of inspection on their vessels, as long as it is carried out in a non-discriminatory manner, chiefly based on the size of the fishing fleet and time spent in the regulatory area (see Control and Enforcement Scheme, article 15).

If fishing fleet vessels should be involved in serious crimes as described in this chapter, this random right to visit embodies an enhanced possibility to uncover them.[16] However, perhaps even more important is that this right to inspection gives comprehensive knowledge about who is operating in the area, and hence who is unknown and should be subjected to closer surveillance.

The right to Search, to Seize, and to Arrest

Experts in marine and coastal law might find it strange that the right to search and seize is discussed in the same paragraph. Having my expertise in criminal procedure law and police law, this is natural: On land, you will normally be authorized to seize what you are searching. As shown below, this is not always so at sea, maybe not being even the principal rule.

From UNCLOS article 110, paragraph 2, it follows that after boarding a vessel suspected of engaging in piracy or the slave trade, the coastal state may proceed to a further examination on board the ship if the suspicion remains. Nevertheless, after the search has been carried out, the right to seize differs between piracy and the slave trade. In UNCLOS article 105 every state may seize a pirate ship on the high sea. For ships where suspicion of engagement in the slave trade is strengthened after boarding and searching, no such right is conferred. As stated in (Churchill and Lowe, 1999), '... slave trading is not in international law analogous to piracy (...): only the flag State may proceed to seize the ship or arrest those on board if the suspicion that the ship is engaged in slavery is well-founded. The other States may only report their findings to the flag State'.

As shown above, the Protocol to Prevent, Suppress and Punish Trafficking in Persons, Especially Women and Children contains no regulation of the possible right of a coastal state to board a foreign vessel, if suspected of being engaged in trafficking in persons; hence there is no right to search or seize. But if the coastal state has reasonable grounds to suspect that a foreign vessel is engaged in what is generally considered to be a less severe crime, namely the smuggling of migrants by sea, it can notify the flag state, and request confirmation of registry and, if confirmed, request authorization from the flag state to take appropriate measures with regard to that vessel. The flag state may authorize the requesting state:

a   To board the vessel, but also
b   To search the vessel, and
c   If evidence is found that the vessel is engaged in the smuggling of migrants by sea, to take appropriate measures with respect to the vessel and persons and cargo on board, as authorized by the Flag state.

> The Protocol against the Smuggling of Migrants by Land, Sea and Air, article 8, paragraphs 2b and 2c

It follows that even if a right to board is authorized, it is up to the discretion of the flag state to decide whether the coastal state should be allowed to search the ship and what to do with persons, cargo, and vessel if the suspicion is supported. It might be more than a theoretical possibility that the flag state has no interest in seeing the migrants returned but is more than happy to see them go. If so, the coastal states can do little before the vessel enters a contiguous zone.

If the coastal state has reasonable grounds to suspect the vessel to be involved in a terrorist activity covered by the SUA Convention and Protocol, or in the illicit trafficking of narcotic drugs, the procedure is more or less the same as for smuggling of migrants.

It follows from the SUA Protocol, article 8bis, paragraph 5b that the flag state has to authorize searching in the same way as boarding. If evidence of terrorist conduct is found, the flag state may authorize the requesting party 'to detain the ship, cargo and person on board pending receipt of disposition instructions from the flag state' (see SUA Protocol, article 8bis, paragraph 6). Both authorizations may be subjected to conditions, including obtaining additional information from the requesting coastal state (see SUA Protocol, article 8bis, paragraph 7). If the coastal state wants to take the policing actions further than searching, the preference in favour of the flag state becomes even more pronounced. SUA Protocol, article 8bis, paragraph 8 states:

> For all boardings pursuant to this article, the flag State has the right to exercise jurisdiction over a detained ship, cargo or other items and persons on board, including seizure, forfeiture, arrest and prosecution. However, the flag State may, subject to its constitution and laws, consent to the exercise of jurisdiction by another State having jurisdiction under article 6.

According to the UN Narcotic Drugs Convention of 1988, article 17, paragraphs 4b and 4c, the flag state can authorize the requesting state to search the vessel, and if evidence of involvement in illicit traffic is found, take appropriate action with respect to the vessel, persons, and cargo on board. If both the requesting state and the flag state are parties to the Council of Europe, the UN Convention can be supplemented by the Council of Europe Agreement on Illicit Traffic by Sea of 1995, but this does not provide the coastal state with significantly more comprehensive police powers.

As mentioned above when discussing the contiguous zone, UNCLOS, article 33, paragraph 1a states: '... the coastal State may exercise the control necessary to: prevent infringement of its customs, fiscal, immigration or sanitary laws and regulations within its territory or territorial sea...'.

The coastal state has rights over foreign-flagged vessels in the contiguous zone when there is a reasonable suspicion that the vessel is involved in all forms of smuggling, illegal immigration, or carrying substances that might be threatening to health, for example, nuclear, biological, and chemical substances. How comprehensive these rights are seems to be disputed. Some argue that the preventive character of the paragraph limits it to inspections and warnings. In contrast, others argue that the coastal state can carry out legal measures like searching, seizure, and arresting the vessel and escorting it to port as well (see Klein, 2011 for more details). Like Klein, I agree with this more progressive approach, but also that a proportional clause must restrict the rights of the coastal state.

Conclusion

The coastal states covered by this article are Russia, Norway, Iceland, and Denmark (Greenland). The possibilities of these coastal states to police the Arctic Ocean seem to be reasonably satisfactory in their territorial waters and the contiguous zone. If spotted, a suspicious vessel heading for the coastal state can be interdicted and searched 24 nautical miles (approximately 45 km) from shore. If it is considered necessary or desirable to control the vessel at an earlier stage, all legal instruments require that the coastal state asks for authorization from the flag state. The only exceptions are stateless vessels (not flying any flag) or vessels engaged in piracy, according to the stipulations in UNCLOS, article 101.

If one of these four coastal states wishes to police the high seas of the Arctic Ocean, they do not have to wait only for the response from the flag state, but also that this response is responsible. Relying on the flag states to see the request as promoting a beneficial purpose, namely enhancing maritime security, and as such benefiting all states, is an attractive, but probably not an entirely realistic thought. In situations like the hypothetical scenario which have been the framework for this paper, one might not receive an answer at all.

So, what is called for? It is not realistic that the international maritime community will agree on a regional treaty for policing the Arctic Ocean. The best way is probably for the community of democratic, stable, and peaceful coastal states, which are not suspected of any hidden agendas, to work towards amendments of existing multinational treaties, in order to create a balance between the seriousness of the suspected crime and the policing powers. In this work, it would be helpful if the leading academics in the field could suggest amendments and additional rules that are needed.

Notes

1 The idea for this article, originates from participation in 'MARPART – Maritime preparedness and international partnership in the high north'. The MARPART programme focused mainly on safety and the rescue limitations caused by the climate and available resources, and the need for further research, which derived from that. My contribution was to outline the existing legal possibilities to prevent and combat possible man-shaped security treats in the future. As this at present is a non-existing problem, it contains no empirical research. To see how the MARPART programme has influenced how the Arctic Ocean is defined and which geographical parts are discussed in the chapter, see 2.1.1 of the programme.

2 See http://e24.no/energi/shipping/russisk-tankskip-med-norsk-gass-satte-rekord/24125864

3 For obvious reasons, the USA was a key participant in these negotiations. This participation might have had a chilling effect on the willingness to reach a consensus: In allowing US warships to intercept flag state vessels on the high seas or take forceful actions in the coastal state territorial water, many countries might have feared that the SUA convention could turn in to a US convention.

4 www.neafc.org/page/27

5 The reason for this geographical limitation is that this paper originates from participating in the MARPART programme (see footnote 1).
6 The obligation to prosecute or extradite, which is important because it sends a message to the criminals that their acts will not go unpunished, will not be discussed. This obligation is more from law enforcement than from policing.
7 The question is only relevant for ships. For aeroplanes and helicopters, there is no 'innocent passage'. Flying over territorial waters will be a violation of the sovereignty of the state, see page 148 in Ruud and Ulfstein: Innføring i Folkerett (2011) Oslo, Universitetsforl.
8 Global Maritime Distress and Safety System (GMDSS) sea areas serve two purposes: to describe areas where GMDSS services are available and to define what radio equipment GMDSS ships must carry (carriage requirements). The number and type of radio safety equipment ships have to carry depends upon the GMDSS areas in which they travel. GMDSS sea areas are classified into four areas: area 1, area 2, area 3 and area 4. https://en.wikipedia.org/wiki/Global_Maritime_Distress_and_Safety_System.
9 Vessels not in the innocent passage must obey all legislation of the coastal state, see page 95 in Churchill, R. and A. V. Lowe (1999). The Law of the Sea. Manchester: Manchester University Press.
10 Slave trade will be discussed below in 'The Slave Trade, Trafficking in Persons and Smuggling of Illegal Immigrants', but unauthorized broadcasting will - appropriately - be left in silence.
11 It is not difficult to agree with Guilfoyle when he says: "Viewed in anything but a historical light the law of the sea appears inconsistent – if not morally incoherent – in allocating boarding rights and enforcement jurisdiction." Guilfoyle, D. (2009). Shipping interdiction and the law of the sea. Cambridge, Cambridge University Press.
12 . In 1985, members of The Palestinian Liberation Front entered the Italian ship, Acillo Lauro, at the port, and hijacked it at high sea in the Mediterranean Sea.
13 According to Klein Klein, N. (2011). Maritime security and the law of the sea. Oxford, Oxford University Press. the agreement is described as intimately connected to the UN 1988 Drug convention.
14 In Norwegian, the first is called "The loop ocean" and second "The loophole".
15 A convention which is empowered by United Nations Agreement for the Implementation of the Provisions of The United Nations Convention on the Law of the Sea of 10 December 1982 relating to the Conservation and Management for straddling Fish Stocks and highly migratory Fish Stocks from 1985, chiefly by its Art. 21.
16 As shown in Roach, J. A. (2004). "Initiatives to enhance maritime security at sea." Marine Policy **28**: 25, this is more than a hypothetical possibility

References

Byers, M. and J. Baker. (2013). *International Law and the Arctic*. Cambridge:Cambridge University Press.

Churchill, R. and A.V. Lowe (1999). *The Law of the Sea*. Manchester: Manchester University Press.

Frostad, M. (2016). *Voldelige hav: pirateri og jus*. Oslo: Cappelen Damm Akademisk.

Guilfoyle, D. (2009). *Shipping Interdiction and the Law of the Sea*. Cambridge: Cambridge University Press.

Guilfoyle, D. (2013). *Piracy off Somalia and counter-piracy efforts. Modern piracy: legal challenges and responses*. Cheltenham: Edward Elgar: 35–60.

Guilfoyle, D. (2014). Piracy and terrorism. In A. Skordas and P. Koutrakos, *The Law and Practice of Piracy at Sea: European and International Perspectives.* Oxford: Hart: 33–52.

Gundhus, H.I. et al. (2007). *Polisiær virksomhet: hva er det - hvem gjør det?: Forskningskonferansen 2007.* Oslo: Politihøgskolen.

Jesus, J.L. (2003). Protection of Foreign Ships against Piracy and Terrorism at Sea: Legal Aspects. *The International Journal of Marine and Coastal Law,* 18(3): 37.

Klein, N. (2011). *Maritime Security and the Law of the Sea.* Oxford: Oxford University Press.

Marchenko, N.A., Borch, O.J., Markov, S.V., Andreassen, N. (2015). Maritime Activity in the high north - The range of unwanted incidents and risk patterns. Paper presented at the Paper presented at the 23rd International Conference on Port and Ocean Engineering under Arctic Conditions, 14–18 June 2015.

Mawby, R.I. (2007). Plural Policing in Europe. In H.I. Gundhus, P. Larsson and T.-G. Myhrer, *Polisiær virksomhet: hva er det - hvem gjør det?.* Oslo: Politihøgskolen: 7.

Roach, J.A. (2004). *"Initiatives to enhance maritime security at sea."* Marine Policy **28**: 25.

Roth, M.P. (2014). Historical Overview of Transnational Crime. In P. Reichel and J.S. Albanese, *Handbook of Transnational Crime and Justice.* Los Angeles, CA, Sage Publications: 5–22.

Ruud, M. and G. Ulfstein (2011). *Innføring i folkerett.* Oslo: Universitetsforl.

8   Coast Guard Co-operation in the Arctic
A Key Piece of the Puzzle

Rebecca Pincus[1]

Introduction

In October 2015 the eight Arctic states launched the Arctic Coast Guard Forum (ACGF), an international co-operative effort that brings together the coast guards of Canada, Denmark, Finland, Iceland, Norway, Russia, Sweden, and the USA. The service chiefs of the eight coast guards formally launched the ACGF at the US Coast Guard Academy. The creation of the ACGF marks an important evolution in international co-operation in the Arctic region, as well as a step forward in crisis response and emergency management. This chapter will assess the contributions of the ACGF and provide the context for its establishment.

The ACGF serves as a forum for co-operation and joint efforts to address shared objectives in the Arctic region, where harsh conditions, great distances, and the scarcity of infrastructure and assets exerts significant pressure on coast guard organizations. These conditions heighten interest in co-operation as an efficient and cost-effective means of improving mission delivery. Furthermore, significant overlap exists among the Arctic states regarding objectives and threats, and there is little meaningful conflict. The long history of co-operation among these eight states within the Arctic Council facilitates ACGF collaboration as well.

The ACGF is not the first or only Coast Guard Forum. Others exist and predate the ACGF: the North Pacific Coast Guard Forum, established in 2000, and the North Atlantic Coast Guard Forum, established in 2007. In addition, the Black Sea Littoral States Border/Coast Guard Co-operation Forum was formally established in 2006 following six years of meetings[2]. The North Atlantic Coast Guard Forum initially grew out of co-operation among Arctic states: Denmark, Finland, Iceland, Norway, and Sweden, but soon grew to include 17 additional European countries, as well as the USA and Canada.[3] In 2009 the Icelandic Minister of Justice, Ragna Árnadóttir, described changing Arctic conditions as an argument for the creation of the North Atlantic Coast Guard Forum: 'The changing conditions in the Arctic region, and their implications for maritime security in the North Atlantic... the opening up of new sea routes' and other emerging maritime challenges meant that 'we are forced to deal with a range of significant new issues, many of them relating to security'.[4]

Coast guard forums grew out of a problem of co-ordination. Maritime commerce connects coastal nations, and the sea is a great connector. Coast guards around the world function to protect human life and ensure law and order at sea. In maritime areas beyond national jurisdiction, like the high seas, and in adjacent sovereign waters, co-ordinating the activities of different national coast guards creates complex co-ordination problems. The ease with which vessels can cross jurisdictional boundaries—for example, by crossing from one state's exclusive economic zone (EEZ) to another state's, or from an EEZ to the high seas, means that coast guards of adjacent states could benefit from co-operation in order to counter illegal maritime activity. Onshore, a criminal fleeing the police could be stopped at a border crossing. If he manages to cross the border, the police could then call their counterparts in the neighbouring state. At sea, there are no border crossings, raising the value of cross-border co-operation. In law-enforcement activities at sea, coast guards could be required to provide evidence in support of prosecution. These processes may also involve co-operating with neighbouring authorities. In maritime border areas where resources are concentrated—particularly fisheries, as fish are mobile resources—cross-border coast guard co-operation could be vital to ensuring peaceful development and avoiding conflict.

In addition, the co-ordination of coast guard activities, many of which revolve around shared objectives like human safety, offers efficiency benefits. There are also strong cultural norms of co-operation and mutual assistance among the community of mariners, and coast guard co-operation can be further appreciated in this context. One of the world's strongest maritime laws, the International Convention for the Safety of Life at Sea (SOLAS), specifically obligates ships, 'on receiving a signal from any source that a ship or aircraft or survival craft thereof is in distress, is bound to proceed with all speed to the assistance of the persons in distress'.[5] While this language reflects a long-standing mariner norm of mutual aid, the SOLAS Convention also specifically mandates that coastal states provide 'to ensure that any necessary arrangements are made for coast watching and for the rescue of persons in distress at sea round its coasts'.[6] The layers of protection formed by a culture of mutual aid, as well as state-provided rescue services, create strong norms that unite mariners in crisis, whether public or private citizens. Structuring interactions through coast guard networks, as well as other networks and organizations relating to maritime activity, provides predictability and reduces the chance of friction in crisis.

This is particularly important in the international context, which is generally characterized by the absence of strong co-ordinating institutions. Therefore, a networked approach is common and valuable. Coast guard fora can be understood as networks of coast guards, which also facilitate connections to additional soft-security organizations at the national level.

This chapter will explain the Coast Guard Forum concept, with particular focus on the Arctic Coast Guard Forum as an example of successful regional networked soft-security practice.

Theory

Network theory provides an important framework for analyzing and understanding the ACGF, as well as the other coast guard fora around the world. The study of networks connecting a wide variety of organizations has provided a rich literature and useful insights. Much of the research in this field has focused on the provision of public services via networks, including community health, emergency services, and crime prevention.[7] Extending this research to public services in the maritime domain is therefore a natural adjunct to a rich field of study. While the study of networks in public policy has yielded a diverse mix of concepts, definitions, and models, some overall consistencies stand out. As Adam and Kriesi (2007) observe, the image of a network is intuitive: 'regular communication and frequent exchange of information lead to the establishment of stable relationships between actors and to the coordination of their mutual interests'.[8] Similarly, Brass et al. (2004) define a network 'as a set of nodes and the set of ties representing some relationship, or lack of relationship, between the nodes', for example, as found in alliances, collaborations, work (goods and services), or boards.[9]

As Provan and Milward (1995) argued, networks must be studied at multiple levels to fully understand the motivations of the organizations participating, the impact to the community, and the functioning of the network itself.[10] Taking a broader perspective, networks function in the international context as well, which can be considered the community of states. In recent decades, much scholarly attention has been paid to the European Union as a transnational policy network linking multiple types of institutions at multiple levels in the international context. For states and coast guards alike, participation in a network, like the ACGF, provides benefits (and also imposes costs) at the international, national, and organizational levels. Network theory provides useful models for analyzing and understanding networks, their ways of functioning, and their benefits and costs.

The analysis model used here draws on the classic framework established by Provan and Milward (1995). They note that any analysis of network effectiveness must take into consideration the perspectives of three categories of network constituents: the principals who oversee the network, the agents who participate in the network, and the clients who receive the services of agents (and advocate to the principals).[11] Each of these three groups can perceive benefits and costs differently. This chapter assesses the work of the ACGF as a network along these three levels, using an adapted version of this framework. Here, we will analyze the effectiveness of the ACGF as a network by considering the perspectives of the states that created the ACGF; the coast guards that participate in it; and the communities that are served by coast guards. As Provan and Milward (op. cit.) pointed out, different network constituents have differing criteria for measuring effectiveness. While the communities served by the network focus on whether or not the problem is being solved (the lowest or most operational level), the states or principals that oversee the network are primarily focused on cost, public perceptions, and indicators or data relating to problem incidence (highest or most abstract level). In the middle are the actors or coast guards, who must focus on community outcomes, principal evaluation, and agency impacts (good/bad). These levels of analysis will provide a useful theoretical basis for understanding the ACGF as a network, as well as other international maritime networks.

Methodology

This chapter will explain the functions of coast guards and interpret coast guard organizations, in particular the ACGF, as networks. Information about the ACGF is drawn from publicly reported news articles, as well as the author's first-hand involvement with the ACGF in 2015–18. This involvement permitted participant observation, informal interviews, and document review during the key early years of the ACGF.

The establishment of the ACGF and its activities and accomplishments provide data for a network theoretical analysis. The diversity of coast guard organizations participating in the ACGF provides another basis for analysis. Not only do coast guard fora act as networks to address the co-ordination problems posed by soft security problems in the maritime environment, but coast guards themselves vary significantly. As a result, incorporating coast guard subjects into network theoretical studies could provide a rich basis for analysis. Organizations join and participate in networks in order to gain benefits, and therefore it is important to consider what benefits flow from coast guard network activity. Participation in the ACGF yields benefits at both the micro and macro level.

Results and Discussion

This chapter explores the important role of coast guards in emergency and disaster response in the Arctic. Working together, albeit in a loose manner, coast guards have formed regional networks like the ACGF. The literature on networks makes clear that it is appropriate to consider the ACGF and other regional coast guard fora as networks: as Klijn and Koppenjan (2000) observe, networks are simply enduring patterns of interaction between actors, which 'emerge around policy problems and resource clusters' and, over time, these patterns of interaction create rules.[12] The ACGF and other regional coast guard fora are formalized systems for interaction among coast guards around common policy problems.

What Can be Learned By Spplying Network Theory to the Arctic Coast Guard Forum?

First, it is important to recognize the high diversity of coast guard organizations. Coast guards play an important role in achieving national objectives in the maritime zone, but those objectives differ from state to state. While exact roles and authorities differ, coast guards generally perform a variety of policing functions, including the protection of fisheries, control of maritime borders and boundaries, interdiction of illegal activity, search and rescue (SAR), and oversight of and response to offshore industrial activity. Furthermore, there is high diversity in coast guard structures, placement within national structures of government, and types of authorities.

Many of the problems that coast guards combat against are transnational in nature and therefore highly amenable to joint co-operative efforts. For example, marine oil spills are universally undesired and can flow across national borders or affect regionally important marine resources like spawning grounds. As a result, neighbouring and regional states are strongly motivated to seek regional co-operative measures on shared priorities, such as preventing oil spills or the interdiction of maritime crime. In the Arctic region, this is true. As a region with low levels of conflict and largely settled national borders, threats to human safety and environmental integrity predominate.

Establishment of the Arctic Coast Guard Forum

The ACGF website explains that it is 'an independent, informal, operationally-driven organization, not bound by treaty, to foster safe, secure, and environmentally responsible maritime activity in the Arctic' (www.arcticcoastguardforum.com).

Unlike other Arctic or maritime fora like the Arctic Council or the International Maritime Organization (IMO), the ACGF is a meeting of operators. While political issues could occasionally arise – for example, the first meeting of the ACGF, which was hosted by Canada in 2014, disinvited the Russian delegation – overall the focus of the forum is on practical co-operation. In this, the ACGF could serve a useful role as a consistent voice for longer-term strategic consistency on Arctic maritime issues.[13] Particularly in the absence of an active and all-inclusive security-focused forum, the ACGF plays an important role in facilitating dialogue and co-operation on softer security issues that can provide a robust basis for discussion.

ACGF Activities and Accomplishments

The ACGF has held annual summits during which the service chiefs have gathered to sign formal statements and decisions. In this chapter, we build upon the meetings of October 2015, June 2016, and March 2017. These meetings, at approximately nine-month intervals, reflect the rapid pace of development of the ACGF. Expert-level meetings have occurred even more frequently, at which staff-level work is done to advance initiatives for summit acceptance. Generally, the ACGF holds both a summit and an experts' working meeting each year, which are hosted by the county that is chairing the secretariat. The USA held the first chairmanship, during 2015–17, and Finland was the chair in 2017–19. In 2019 the chair rotates to Iceland, and in 2021 it will pass to Russia. The chair of the ACGF rotates in lockstep with the chair of the Arctic Council.

The ACGF held its first tabletop exercise (TTX) in the autumn of 2016, and the first live exercise (MMEX), 'Arctic Guardian' in September 2017. The Arctic Guardian exercise marked an important step forward for the ACGF, demonstrating that it could successfully bring together the Arctic states for a multiday live exercise involving five active ships plus a variety of aircraft and smaller platforms. In testing a SAR scenario, the exercise focused on an area of common interest, in which co-operative activity requires a high degree of co-ordination. In order to effectively organize the efforts of multiple ships and planes and to successfully complete a SAR mission while maintaining the safety and security of the personnel and platforms responding, careful attention must be paid to information management, communications, and operating procedures. In a multilateral response, these challenges are heightened. The successful execution of the Arctic Guardian exercise therefore demonstrated that the co-operative work of the ACGF, including its Voluntary Guidelines for Combined Operations, provides a blueprint for future joint missions. The ACGF plans to continue to exercise joint scenarios and refine combined operations in the future.

Thus far, the ACGF has adopted a set of voluntary guidelines for combined operations, a strategic roadmap, a process guide, and terms of reference. The ACGF has identified ten strategic goals:

- Strengthen multilateral co-operation and co-ordination within the Arctic maritime domain, and existing and future multilateral agreements
- Seek common solutions to maritime issues related to the agencies fulfilling the functions of coast guards within the region
- Collaborate with the Arctic Council through the sharing of information
- Facilitate safe and secure maritime activity in the Arctic region, with sustainable development to be promoted as appropriate
- Contribute to a stable, predictable, and transparent maritime environment
- Build a common operational picture to ensure proper protocols for emergency response co-ordination, and safe navigation
- Work collaboratively to advance the protection of the marine environment
- Maximize the potential for Arctic maritime activities to positively impact the communities, lives, and culture of Arctic communities including indigenous peoples
- Integrate scientific research in support of coast guard operations as appropriate

- Support high standards of operations and sustainable activities in the Arctic through the sharing of information, including best practices and technological solutions to address threats and risks (arcticcoastguardforum.com)

Benefits of ACGF Participation

The ACGF brings together a diverse group of coast guards, reflecting the diversity of coast guards found globally. Benefits of ACGF participation can be identified at the levels suggested by Provan and Milward: the principal level, or highest levels of national government that oversee coast guards; the agent level (coast guards themselves); and the client level, or the public safety interest. Despite the strong diversity present in coast guard organizations, they all sit within the principal-client matrix of interests.

Some of the coast guard organizations participating in the ACGF are military, and some are civilian, as can be seen from the ACGF website. Others are quasi-military. They have varying roles, responsibilities, and authorities.

Nevertheless, these eight organizations have been identified by their national governments as having coast guard functional responsibilities in the Arctic region. While the specific functional responsibilities of these eight coast guards vary, as noted above, there are some commonalities. For example, all have been identified by their national governments as having Arctic SAR responsibilities. When the eight Arctic states signed a joint treaty in 2011 on search and rescue response, they bound themselves formally to co-operation in SAR. The treaty, formally titled the 'Agreement on Co-operation on Aeronautical and Maritime Search and Rescue in the Arctic', was signed in Nuuk, Greenland.[14] Treaty text includes language 'recognizing the great importance of co-operation among the Parties in conducting search and rescue operations' and 'emphasizing the usefulness of exchanging information and experience in the field of search and rescue and of conducting joint training and exercises' among other statements. These examples illustrate that the language of the treaty clearly envisioned some means of implementing

Table 8.1 Coast guard organizations participating in the ACGF and their authorities

Canada	Canadian Coast Guard	Fisheries and Oceans Canada
Denmark	Joint Arctic Command	Ministry of Defence
Iceland	Icelandic Coast Guard	Minister of Justice
Finland	Finnish Border Guard	Ministry of the Interior
Norway	Norwegian Coast Guard	Royal Norwegian Navy/Ministry of Defence
Russia	Russian Coast Guard	Federal Security Service
Sweden	Swedish Coast Guard	Ministry of Justice
USA	US Coast Guard	Department of Homeland Security

co-operative measures, information exchange, and joint practice. Like the ACGF, this language is specific to operations and the practical level of SAR. It emphasizes co-operation in conducting operations, as well as information exchange and training relevant to the field of SAR. The close tailoring of the treaty language here points towards co-operation among the practitioners specifically of aeronautical and maritime (distant) search and rescue: coast guard agencies.

In addition, the representatives to the ACGF have responsibility for enforcing fisheries regulations. Improving co-ordination of at-sea fisheries enforcement will advance goals shared by all the Arctic states and is likely to be an important element of ACGF work. In 2015 the five Arctic coastal states signed a joint pledge to establish a moratorium on fishing in the central Arctic Ocean. The 'Declaration Concerning the Prevention of Unregulated High Seas Fishing in the Central Arctic Ocean' was signed by representatives of Canada, Denmark, Norway, Russia, and the USA[15] While not a legally binding treaty, this declaration clearly communicated a joint intention to protect the Arctic 'donut hole' or high seas area surrounded by the EEZs of the five coastal states, from fishing until a regional fisheries management organization is established and greater clarity is reached about the feasibility of emerging fisheries in the central Arctic. The joint resolution pledged that, in order to ensure compliance with the moratorium, the five states would co-ordinate 'monitoring, control and surveillance activities in this area'. These three activities all generally fall within the mission sets of coast guard organizations. Subsequent to the 2015 resolution, a larger group of Arctic and non-Arctic states met during 2015–17 to hammer out terms for a binding treaty along the same lines, which is currently in a draft format and under review.[16] The 'Agreement to Prevent Unregulated High Seas Fisheries in the Central Arctic Ocean' was negotiated by Canada, China, Denmark, the European Union, Iceland, Japan, Korea, Norway, Russia, and the USA. Once this formal treaty has been signed and entered into force, discussions can commence at the ACGF on how to co-operate on collective efforts to enforce the moratorium in a high seas environment.

In 2016, the 'Agreement on Co-operation on Marine Oil Pollution Preparedness and Response in the Arctic' entered into force. This treaty, which binds parties to 'strengthen cooperation, coordination and mutual assistance...in order to protect the marine environment from pollution by oil' also bears on ACGF activities. While not all parties to the ACGF have oil spill response missions, most of them do, and all of them have operational responsibilities that would include response to any emergency situation in their waters. For example, while Finland, Iceland, and Russia do not include their coast guards among the designated assistance authorities or competent national authorities (the first line of response authorities formally identified in the treaty), the coast guards of Finland and Iceland identify maritime pollution prevention and response among their mission sets[17], and the Russian coast guard's role in co-ordinating federal control of natural resources and combating maritime crime would seem to indicate that it would be involved in maritime pollution response in some capacity. The ACGF itself has indicated that marine environmental response will be an increasing focus of activity in coming years.[18]

Beyond this, many of these organizations manage border controls at sea. In May 2018 the latest Arctic treaty went into effect. The 'Agreement on Enhancing International Arctic Scientific Co-operation' was intended to 'increase effectiveness and efficiency in the development of scientific knowledge about the Arctic'. While this may not seem of immediate relevance to coast guards, access to Arctic waters has occasionally been difficult for Arctic marine research. Dr Farrell of the US Arctic Research Commission explained that barriers to research often included 'denied visas, the inability to carry equipment and samples across national borders, or denial of access to data'.[19] As agencies that control maritime borders, the ACGF members will be implicated in the implementation of this scientific cooperation treaty, although as Dr Farrell notes, given the softness of the language in the treaty, 'the proof will be in the pudding' of the treaty's effectiveness and impact.

As the preceding review of recent Arctic treaties illustrates, the states of the region will depend on their coast guards for the implementation of their treaty obligations and fulfilment of shared objectives. The implementation of international agreements is facilitated by the ACGF and therefore the principals, or national governments of the ACGF, benefit from the network functioning to improve implementation and possibly improve efficiency (lower costs). Coast guards will serve as either the leading or key support agencies across a variety of missions, including SAR, environmental response, border control, and fisheries enforcement. These treaties contain explicit commitments to improving co-ordination and information exchange, and thus the creation of the ACGF is an important step in the successful execution of international intentions regarding Arctic protection and management.

Of course, many other international treaties apply to the Arctic region and involve coast guard agencies in their execution. A full review is beyond the scope of this chapter. Nevertheless, even the examples discussed here make clear that the ACGF, and the individual coast guards that comprise the Forum, play a critical role and serve as important action agents in fulfilling diplomatic agreements. In this, the ACGF works closely with the Emergency Prevention, Preparedness, and Response (EPPR) working group of the Arctic Council. EPPR is not an operational organization, but rather co-ordinates projects aimed at strategic and capacity issues relating to Arctic emergencies. EPPR collects and shares data and develops guidance and risk assessment methodologies; co-ordinates training; and exchanges information.[20]

At the more distributed or operational level, there is even more robust co-operation among Arctic coast guards. The ACGF brings together coast guard chiefs of service, and the experts who sit on the working groups are generally drawn from senior staffs in relevant offices from headquarters in capital cities. However, on-scene co-operation between Arctic coast guards generally occurs in border areas, between frontline district units, and at far more junior levels. One could envision a disconnect between the amicable joint statements and group photos that emanate from service chiefs' summits and the actual conduct of missions at sea. Coast guard officers and enlisted are formidable men and women, who are accustomed to hard-charging action in dangerous and unpleasant conditions. However, examples of recent emergency responses involving bilateral or multilateral co-operation among ACGF members indicate that the spirit of co-operation that is so visible in ACGF achievements is also present on the frontlines and contact zones between these organizations. These operational benefits make clear that ACGF participation benefits the coast guard organizations—the network actors—themselves.

For example, Norway and Russia, which share a maritime boundary in the Barents Sea that was only settled in 2010, teamed up in June 2018 for their annual joint SAR/oil response exercise – this after each country had just wrapped up military exercises largely aimed at each other (and NATO).[21] However, for the annual Exercise Barents, the Norwegian Coast Guard and the Russian Navy and Coast Guard sent a variety of afloat and air assets to participate in the scenario, which simulated the sinking of a fishing ship along the maritime border. This scenario is also realistic. In 2007 Russia requested Norwegian assistance to evacuate the crew of a cargo vessel that was sinking. As exercise participants noted, regularly practicing scenarios and exercises improves the effectiveness of joint responses in times of real crisis.

As long-standing allies, Canada and the USA also co-operate closely in coast guard activities, and their service chiefs meet together for the Canada-United States Coast Guard Summit annually. In 2017 the US Coast Guard hosted the meeting in Grand Haven, Michigan. The two service chiefs signed an update to the Canada-United States Joint Marine Pollution Contingency Plan, which 'promotes a coordinated system for planning, preparing, and responding to harmful substance incidents in adjacent waters' and facilitates the close coordination between the two services.[22] A CANUSNORTH Annex provides specific guidance for waters in the Beaufort Sea.[23] In the Arctic, the two regional components that execute the Joint Contingency Plan (JCP) are Alaska's District 17 and the Central and Arctic Region of the Canadian Coast Guard. At the summit, Arctic-specific issues also starred on the agenda, including a discussion of co-operation through pan-Arctic fora.[24]

In the Arctic, the harsh conditions and scarcity of infrastructure and assets have even encouraged co-operation between the USA and Russia, despite political tension between the two states, which share a maritime boundary in the Bering Sea. In 2011 the USA and Russia signed a JCP to combat pollution in the Bering and Chukchi Seas and executed a memorandum of understanding with the State Pollution Control, Salvage, and Rescue Administration.[25] In November 2018 the US Coast Guard conducted a joint marine pollution contingency seminar and table top exercise with Russia.[26] Exercises and enabling legal mechanisms to facilitate engagement are important in the Bering region, where harsh conditions and dangerous weather raise the stakes for marine activity. The frequency of maritime incidents in the Bering makes US-Russia cooperation a virtual necessity to protect human life and environmental quality.

For example, in January 2018 the Russian Coast Guard alerted Alaska's US Coast Guard District 17 to a possible illegal, unreported, and unregulated (IUU) vessel fishing in the US EEZ.[27] The Russians were tracking the vessel, the *Sea Breeze*, and suspected it of illegally fishing for crab in the Bering Sea and of links to organized crime. In response, the US Coast Guard launched a search plane from Air Station Kodiak and located the vessel, which failed to respond to hail and query from the plane. The US Coast Guard passed all its information back to the Russian side, which deployed ships to board and seize the *Sea Breeze*. The Kamchatka Border Guard/Coast Guard unit cited the Sea Breeze for fishing in the Russian EEZ without authorization and for failure to maintain a logbook and escorted the vessel back to Petropavlovsk-Kamchatsky.[28]

In 2014 the sinking of the South Korean fishing vessel *Oryong 501* demonstrated co-operation between the US Coast Guard and Russian authorities, in this case the Maritime Rescue Service of the Federal Marine and River Transport Agency.[29] The *Oryong 501* sank in the Bering Sea when a large wave struck the vessel while it was hauling in nets loaded with pollock. Five other fishing vessels in the area responded to the master's distress call, and rescued seven individuals, one of whom was reported to be a Russian fishing inspector.[30] The vessel is reported to have sunk in Anadyr Bay, and over 50 remaining people were lost. According to the US Coast Guard, watch standers in Juneau, Alaska received notice from the *Oryong*'s emergency locator beacon and immediately contacted the Russian rescue co-ordination centre in Vladivostok, as the ship was in Russian waters.[31] Russian SAR mission co-ordinators actively supported the search, which involved US Coast Guard and Air Force assets along with South Korean planes, ships, and personnel. In addition, five Russian fishing vessels participated in the search: the *Karolina-77*, the *Zaliv Zabiyaka*, the Pelageal, the Astronom, and the Vladimir Bradyuk. While the Russian Marine Rescue Service appeared to be in the lead for the response effort, the Russian Border Guard seemed to be monitoring the arrival of South Korean search vessels. A Russian Antonov AN-26 airplane was described by Interfax as 'ready' to join the search from Anadyr.[32]

These examples illustrate the depth of coast guard co-operation in the Arctic region at a more distributed level. The challenges of Arctic maritime operations make coast guard co-operation complex, but important.

Summary

To summarize, participation in the pan-Arctic ACGF generates important macro-level benefits for all participants:

- Information sharing: including of best practices and new developments to address the challenges of the common Arctic operating environment, including ice, communication difficulties, long distances, extreme weather, subsistence practices, and fragile ecosystems. While the different Arctic areas of responsibility of the eight Arctic states vary, there are common challenges inherent in a cold and stormy climate. Each Arctic state has developed unique skills and experiences that can inform and help its neighbours, improving their abilities and propagating best practices around the region. For example, the ACGF has held workshops on SAR training aimed at improving Arctic-specific SAR training across the ACGF network.

- Efficient progress towards shared objectives: the eight Arctic states confront a shared challenge of protecting the unique and fragile Arctic, while ensuring a smooth transition to a new phase of human presence and activity in the region, all while managing the effects of rapid climate change and continuing to combat legacy pollution. Outside actors are increasingly asserting their role in decision making, and deteriorating relations between Russia and NATO add further complexities to the emerging Arctic geopolitical picture. In response to these trends, the Arctic states have identified a coherent set of principles and goals, articulated through individual national strategies as well as joint statements emanating from the Arctic Council, which will be implemented jointly. The work of the ACGF is crucial to the achievement of common goals, since coast guards embody government at sea, executing the critical roles of enforcements and sovereignty. An example of this progress is the development of an ACGF strategic roadmap, as well as a set of guidelines for combined ACGF operations.
- Relationship building: the development of a community of coast guards in the Arctic region helps build resilience against political developments that may adversely impact co-operation. The imposition of sanctions in response to Russia's annexation of Crimea, and the renewal of NATO exercises in response to increased Russian military activity, could have halted co-operation between the biggest Arctic state and all the others, an outcome that would have been likely to result in greater risk of human and environmental damage. Just as the other Arctic states co-operated with the Soviet Union during the depths of the Cold War, on shared goals like the protection of polar bears and the clean-up of radioactive waste, so today the so-called 'soft security' goals of human safety and environmental protection serve as bridge issues. In times of high tension, these bridge issues can keep channels of communication open, and in times of reduced tension, they can serve as foundations upon which to build more broad and robust dialogue. Within the ACGF, relationships have been built through regular meetings that are held twice yearly. These are supplemented by regular live exercises and tabletop exercises.

Conclusion

Coast guard co-operation in the Arctic operates at two levels. First, a macro-level, pan-Arctic dialogue that is best exemplified by the ACGF, but also exists in the EPPR working group of the Arctic Council, and in tracks within other international fora like the IMO. The ACGF is a stark symbol of the value that Arctic states place on coast guard co-operation. Why else would the US Coast Guard or the Canadian Coast Guard meet regularly with their Finnish or Swedish counterparts, countries with which they share no boundaries, and from which they are separated by thousands of miles? One would be hard pressed to find benefits in terms of joint responses to specific incidents, since the countries' areas of responsibility are so distant. The benefits of co-operation here can be best understood at the principal or state level. While the Canadian and Finnish coast guards might not respond together to a maritime incident, given the distances separating their areas of responsibility, Canada and Finland benefit at a more abstract level by building important relationships that could provide benefits in other areas of foreign policy, and by sharing relevant information about Arctic-specific operations that could benefit each coast guard individually.

In addition, coast guard co-operation in the Arctic operates at a lower, more distributed and local level. In a region where sovereign territories and EEZs are adjacent, individual Arctic states co-operate with their neighbours. The extremely high and well-developed level of co-operation among Nordic and Baltic states is not surprising, given their close ties. In addition, the USA and Canada have a unique and close bond. However, the co-operation seen between Norway and Russia, for example, and the USA and Russia, could surprise outsiders who are not familiar with the sheer difficulty of maritime operations in the Arctic. These operational benefits are clearer to communities served by these coast guards, as well as to coast guard operators themselves who could see immediate practical effects from network participation.

In 1989 the Soviet Union sent the 425-foot Vayadaghubsky oil skimmer to Prince William Sound to assist with clean-up operations following the *Exxon Valdez* spill.[33] The huge Soviet skimmer was given approval to work in Alaskan waters and make port calls, although its pumps eventually clogged with tarry oil congealed with seaweed.[34]

Also in 1989, the Norwegian Coast Guard rescued hundreds of passengers and crew from the sinking *Maxim Gorky*, a Soviet cruise ship, which collided with ice in the Greenland Sea. Evacuees waited for rescue in life rafts and on the ice.[35]

Thirty years ago, coast guard co-operation in the Arctic was alive and well. Since then, the Soviet Union has collapsed, and the Arctic environment has begun to change dramatically. Laws and regulations pertaining to tankers and cruise ships alike have evolved, and we have learned from the two severe Arctic incidents, the *Exxon Valdez* and the *Maxim Gorky*. The coast guards in the Arctic remain critical pieces of the puzzle in preventing and responding to these types of emergencies, and their growing co-operation through the ACGF network is a sign of progress. In order to understand the benefits of coast guard networks like the ACGF—as well as others, including the North Pacific Coast Guard Forum—it is important to parse the benefits of network participation to local communities, to the coast guards themselves, and to the local communities. Each of these stakeholders could have unique criteria of effectiveness. At both higher and lower levels of abstraction or proximity to operations, effectiveness can be perceived differently. Nevertheless, this assessment of the ACGF as a network demonstrates that it is already producing a variety of benefits for all stakeholders. As human activity increases in the Arctic, the lessons of past disasters suggest that turning to the ACGF as a network solution to the emerging problem of emergency response management in the Arctic could be a wise approach for states, coast guards, and communities.

Notes

1 Dr Rebecca Pincus is an assistant professor at the US Naval War College. The views offered here are hers alone and do not represent those of the Naval War College, the US Navy, or the Department of Defense.
2 MIA Border Police of Georgia. 'The leaders of the Black Sea Littoral States Border/Coast Guard Agencies met in Istanbul'. 29 November 2014. http://bpg.gov.ge/en/news/visits/the-leaders-of-the-black-sea-littoral-states-bordercoast-guard-agencies-met-in-istanbul.page.
3 Arctic Portal. (2009). 'North Atlantic Coast Guard Forum Meeting in Iceland'. https://arcticportal.org/yar-features/136-north-atlantic-coast-guard-forum-meeting-in-iceland.
4 Remarks by H.E. Ragna Árnadóttir, 2 October 2009. North Atlantic Coast Guard Forum. www.stjornarradid.is/raduneyti/atvinnuvega-og-nyskopunarraduneytid/fyrri-radherrar/stok-raeda-fyrrum-radherra/2009/09/29/Formal-Opening-Session-of-the-Summit-of-the-North-Atlantic-Coast-Guard-Forum-2009.
5 International Convention for the Safety of Life at Sea (1974), IMO. Chapter V, Regulation 10(a).
6 Ibid., Chapter V, Regulations 15(a).
7 Keith G. Provan and H. Brinton Milward. (1995). A preliminary theory of network effectiveness: a comparative study of four community mental health systems. *Administrative Science Quarterly, 40(1)*: 1–33;
    Christopher J. Koliba, Asim Zia, Russell M. Mills. (2011). Accountability in governance networks: an assessment of public, private, and nonprofit emergency management practices following Hurricane Katrina. *Public Administration Review*, March/April 2011: 210–20; Jörg Raab, Remco S. Mannak, Bart Cambré (2013). Combining structure, governance, and context: a configurational approach to network effectiveness. *Journal of Public Administration Research and Theory, 25(2)*: 479–511.
8 Silke Adam and Hanspeter Kriesi. (2007). The Network Approach. In *Theories of the Policy Process*, 2nd ed. (2007). Paul Sabatier, ed. Colorado: Westview Press.
9 Daniel Brass, Joseph Galaskiewicz, Henrich Greve, and Wenpin Tsai. (2004). Taking stock of networks and organizations: a multilevel perspective. *Academy of Management Journal*, Vol. *47(6)*: 795–817.
10 Keith G. Provan and H. Brinton Milward. (2001). Do networks really work? A framework for evaluating public-sector organizational networks. *Public Administration Review*, Vol. *61(4)*: 414–423.
11 Provan and Milward (2001).
12 Erik-Hans Klijn and Joop F. M. Koppenjan. (2000). Public management and policy networks. *Public Management*, Vol. *2(2)*: 135–158.

13 For further support, see Andreas Østhagen. 2 November 2015. 'The Arctic Coast
    Guard Forum: Big Tasks, Small Solutions'. The Arctic Institute. www.thearctic
    institute.org/arctic-coast-guard-forum-big-tasks.
14 Agreement on Co-operation on Aeronautical and Maritime Search and Rescue in
    the Arctic. (2011). US Department of State. www.state.gov/documents/organization/
    205770.pdf.
15 Declaration Concerning the Prevention of Unregulated High Seas Fishing in the Cen-
    tral Arctic Ocean. (2015). US Department of State. www.regjeringen.no/globalassets/
    departementene/ud/vedlegg/folkerett/declaration-on-arctic-fisheries-16-july-2015.pdf.
16 U.S. Department of State. Bureau of Oceans and International Environmental and
    Scientific Affairs. "Meeting on High Seas Fisheries in the Central Arctic Ocean."
    28 November 2017. https://www.state.gov/e/oes/ocns/fish/regionalorganizations/
    arctic/statements/281792.htm.
17 See self-described roles of Finnish Border Guard at www.arcticcoastguardforum.
    com/member-country/finland and Icelandic Coast Guard at www.arcticcoastgua
    rdforum.com/member-country/iceland.
18 Arctic Coast Guard Forum. (2018). "Joint statement from the Arctic states coast
    guards." www.arcticcoastguardforum.com/news/joint-statement-arctic-states-coast-
    guards.
19 John Farrell. (2017). 'New agreement to enhance international Arctic scientific
    cooperation'. ARCUS. www.arcus.org/witness-the-arctic/2017/10/highlight/1.
20 Arctic Council. Emergency Prevention, Preparedness and Response (EPPR) Work-
    ing Group. https://arctic-council.org/index.php/en/about-us/working-groups/eppr.
21 Thomas Nilsen. (2018). 'After weeks with separate war games, Norway and Russia
    again meet at sea for joint SAR exercise'. *The Barents Observer*. https://thebaren
    tsobserver.com/en/node/4006.
22 Canada-United States Joint Contingency Plan. 2017. http://waves-vagues.dfo-mpo.
    gc.ca/Library/40616733.pdf.
23 US. Coast Guard. (2016). COMDINST M16000.14A, *US Coast Guard Marine
    Environmental Response and Preparedness Manual*.
24 US. Coast Guard. (2017). News release: Top U.S., Canadian Coast Guard leaders
    hold summit in Grand Haven. https://content.govdelivery.com/accounts/USDHSCG/
    bulletins/1aeb082.
25 US Coast Guard. (2016). COMDINST M16000.14A, *US Coast Guard Marine
    Environmental Response and Preparedness Manual*. 15–11.
26 Vice Admiral Charles Ray, US Coast Guard. (2018). Prepared testimony on
    'Maritime Transportation in the Arctic: the U.S. Role' before the House Coast
    Guard and Maritime Transportation Subcommittee. 7 June 2018. p. 6.
27 *The Brig: Alaska Fisheries Enforcement News*. 11 April 2018. http://deckboss-the
    brig.blogspot.com/2018/.
28 Ibid.
29 Reuters. Via gCaptain. 1 December 2014. 'Dozens missing after commercial fish-
    ing trawler sinks in the Bering Sea'. http://gcaptain.com/commercial-fishing-
    trawler-sinks-bering-sea-dozens-missing/. See also 'Russia releases findings on
    stricken South Korean ship'. Tempo.Co 3 December 2014. https://en.tempo.co/
    read/news/2014/12/03/114626226/Russia-Releases-Findings-on-Stricken-South-Korean-
    Ship.
30 Paul Hancock. (2014). '501 ORYONG'. Shipwreck Log. https://shipwrecklog.com/
    log/2014/12/501-oryong.
31 Diana Sherbs. 19 December 2014. 'International search in the Bering'. Coast Guard
    Alaska. http://alaska.coastguard.dodlive.mil/2014/12/international-search-in-the-bering/.
32 Interfax. 4 December 2014. 'Rescuers not expecting to find any more survivors
    from Oryong 501 in Bering Sea'. http://www.interfax.com/newsinf.asp?id=556345.

33 Jeff Berlinger. (1989). 'Spreading oil harms more fishing, more shores, more ani-mals'. UPI; Steve Rinehart. (1989). 'Soviet ship to skim oil near Seward'. *Anchorage Daily News.*
34 National Response Team. (1989). 'The Exxon Valdez Oil Spill: A Report to the President'. Available through EPA; Tom Horton. (1989). 'The Exxon Valdez Oil Spill: Paradise Lost'. *Rolling Stone.*
35 Steve Lohr. (1989). 'All safe in Soviet ship drama'. *The New York Times.*

9 Response – Coast Guard Co-operation in the Arctic

A Key Piece of the Puzzle

Captain Roberto H. Torres

The information contained within this response contains the official view of the US Coast Guard and has been furnished to educate the public regarding Coast Guard interests in the Arctic. This comment is not an endorsement of views or opinions of the book's authors or the underlying material contained within the book.

Dr Pincus's chapter (Chapter 8) provides an accurate review of the Arctic Coast Guard Forum (ACGF) and its critical role in Arctic co-operation. While still nascent, the ACGF's strengths are its lean membership, limited to the coast guards or coast guard-like agencies of only the eight Arctic nations, and its operational focus. A completely independent body, the ACGF operationalizes the maritime policy objectives of the Arctic Council and collaborates with its working groups to align efforts and limit redundancies. The ACGF is a bridge between diplomacy and operations.

The ACGF has a simple organizational structure consisting of two working groups (a Secretariat and a Combined Operations Working Group). Working Group members report to the ACGF Principals – normally the heads of service of the eight coast guard organizations. This lean arrangement enables rapid flow of information to decision-makers, with a similarly easy means for Principals to provide feedback to the working-level experts.

The ACGF operationalizes elements of the US Coast Guard Arctic Strategic Outlook's Lines of Effort: 'Enhance Capability to Operate Effectively in a Dynamic Arctic', 'Strengthen the Rules-Based Order', and 'Innovate and Adapt to Promote Resilience and Prosperity'. The ACGF is a unique, action-oriented maritime governance forum where the US Coast Guard and our peer agencies strengthen relationships, identify lessons learned, share best practices, carry out exercises, conduct combined operations, and co-ordinate emergency response missions.

The semi-annual cycle of ACGF meetings includes an Experts Meeting every autumn and a Summit (with Experts and Principals) every spring. Intercessional work by the Experts informs the discussions at meetings and the recommendations to the Principals, while providing decision space and co-ordination time. ACGF Experts co-ordinate with other agencies in their national framework to conduct Forum business, but only coast guard officials attend ACGF meetings.

In 2017 the ACGF conducted its inaugural live search and rescue (SAR) exercise, Arctic Guardian 2017, off the coast of Iceland. This exercise, led by the ACGF Combined Operations Working Group, demonstrated the unique challenges of operating in the Arctic and reinforced the need for international co-operation in this environmentally sensitive area. With the increase of commercial traffic, discussions between the ACGF Principals and Ambassadors during Arctic Guardian 2017 highlighted the criticality of co-ordination in maritime environmental response and the responsibility to ensure that search and rescue (SAR) resources are prepared to respond.

Earlier this year, Finland hosted the ACGF's second live exercise, Polaris 2019, which bolstered our capability to co-operate with our Arctic partners in a response scenario with live assets and real-time decision-making. Plans were in development for a 2020 tabletop exercise and the ACGF's third live exercise in 2021. These expanded objectives of these next exercises will incorporate elements of both Arctic SAR and Arctic marine environmental response.

The vast, remote Arctic region demands coast guards that are ready to work together to co-ordinate response operations in harsh conditions. The ACGF is the ideal platform for the leaders and operators of the Arctic coast guards to share information, plan and carry out exercises, and identify lessons learned.

In the future, more research is needed to fully embed coast guard organizations into network and organization theory. Specifically, structured interviews should capture data from coast guard operators who participate in the ACGF, as well as other coast guard organizations, in order to gain a better understanding of how these networks function and various aspects of network effectiveness. In addition, further research should be conducted in the Arctic region, among coastal communities supported by coast guards, as well as with policymakers who oversee coast guards. As a relatively new organization, the ACGF provides an excellent opportunity to advance our understanding of networks in the maritime − and Arctic −dimension.

10 International Co-operation – the Way to Success

Jens Peter Holst-Andersen

A Difficult Environment

In 2010 I was commanding a navy patrol ship in the Arctic when the area was hit by a local but very severe storm. Afterward, the search and rescue (SAR) authorities realized that 14 tourists in kayaks were missing and 25 other people were unaccounted for. Anywhere on the planet, this would have been a serious and disastrous situation calling for all SAR resources to be activated immediately. That was also the case in this situation. However, an important difference from more southerly regions was that as the nearest SAR asset, we were approximately 500 nautical miles (about 900 km) away.

The Arctic region presents a very difficult and complex environment when it comes to emergency prevention, preparedness, and response. Remoteness, lack of infrastructure, limited means of communication, and severe weather conditions are just a few examples of the risks and challenges of this unique and extremely sensitive region.

For many years, the governments of the Arctic states did not focus on these challenges, but that has changed recently in response to several trends. Climate change and increased accessibility are the two major drivers. The people living and working in the Arctic have always been aware of these risks and have tried to adapt to them. But with an ever-increasing presence of people and equipment not prepared for the Arctic environment, the risk of accidents could rise. Today, there is a general recognition that the risks need to be addressed and managed through emergency prevention, preparedness, and response.

Cooperation, Co-operation, Co-operation

In real estate, the mantra is that three things are important: location, location, location. When it comes to emergency prevention, preparedness and response in the Arctic, it is: co-operation, co-operation, co-operation. No single government or emergency response authority has – or ever will have – the resources that are needed to fully manage the risks related to disasters in this region. The authorities are very well aware of this. And while the Arctic states know their responsibilities and the risks, they also know that keeping an emergency response regime comparable to what is available in more populated areas would be astronomically costly.

This is where international co-operation comes in handy. Governments and emergency response authorities co-operate worldwide, but in the Arctic region it is an absolute necessity if they are to deal with complex situations like oil spills, large-scale incidents, or mass rescues in this enormous and sparsely populated area. The existing approach to international co-operation has several layers, consisting of a range of bilateral, multilateral, and international/ regional agreements. These agreements can be activated according to the scope of a situation. An example: in the event of a large oil spill in the Icelandic part of the area between Iceland and Greenland, the authorities would initially activate the *bilateral* SOP COOP SAGA agreement between Iceland and Denmark (Denmark is responsible for most coast guard duties around Greenland). If that proved insufficient, Iceland could activate the multilateral Copenhagen Agreement. If the situation expanded still further, then Iceland could activate the international Agreement on Co-operation on Marine Oil Pollution Preparedness and Response in the Arctic (MOSPA agreement). Similar arrangements exist in relation to SAR.

This approach is very rational in the way that it pools resources, but this pooling is also its weakness. None of the agreements is better than the resources that the state authorities put into them. The operationalization is complicated because of national systems, procedures, and legislation. And command, control, and communication during an incident is a potential nightmare. So, is there any hope for effective response?

International Fora

I believe that there is. Continuous dialogue and engagement are necessary to make international co-operation and agreements effective, credible, and relevant. The most logical approach to this is through permanent international cooperative fora.

The Arctic Council Working Group Emergency Prevention, Preparedness and Response (EPPR) is an example of an international forum in which all Arctic states are members, together with the indigenous people of the Arctic (called 'Permanent Participants' in the Arctic Council) and participation from observers from other states and organizations. The EPPR is mandated to contribute to the prevention, preparedness, and response to environmental emergencies, SAR, and both man-made and natural disasters. The EPPR has been given a role by the states in promoting the implementation of two international agreements – the MOSPA agreement and the Agreement on Co-operation on Aeronautical and Maritime Search and Rescue in the Arctic (SAR agreement). The EPPR is not an operational organization, but it works by developing guidance and risk assessment methodologies; co-ordinating response exercises and training; and exchanging information on best practices with regards to prevention, preparedness, and response to accidents and threats from unintentional releases of pollutants and radionuclides, and to the consequences of natural disasters.

Another forum is the Arctic Coast Guard Forum (ACGF), which was established in 2015. In contrast to the EPPR, the ACGF is an operationally oriented forum, and its members are agencies representing the coast guard functions of all eight Arctic states. The ACGF is an independent, informal, operationally driven organization, and its aim is to foster safe, secure, and environmentally responsible maritime activity in the Arctic. The ACGF also works directly with the MOSPA and SAR agreements but from an operational perspective.

These two fora complement each another in several ways in the maritime domain. The EPPR represents the states and Permanent Participants, and it works primarily on a strategic level with a direct link to the political level in the member states. The ACGF represents the operators through coast guards (mostly), but also through other operational national maritime agencies. The EPPR and the ACGF share most of their mandates, but on different levels. And both fora work practising the MOSPA and SAR Agreements. Together, the two fora represent the full spectrum from policy to actual operational assets.

In the long run, this should represent a beautiful partnership. But it is still in its early stages. Co-operation between the two fora is being explored and developed at an increasing pace, and 2020 is the first year with direct and formal co-operation on the planning and execution of exercises related to the MOSPA and SAR agreements. There is huge potential for productive and practical co-operation that needs to be explored and developed.

There are other fora and time-limited international projects (e.g. the Barents Co-operation and the EU Horizon 2020 project, Arctic and North Atlantic Security and Emergency Preparedness Network [ARCSAR]) that work on nearly the same area as the EPPR and the ACGF, but none of them has the full membership of all eight Arctic states.

Organisational Competition

When you work in international co-operation, you quickly learn that excellent co-ordination is both critical and enormously energy-consuming. It is not just within one's own forum or organization, either; co-ordination and de-confliction with related fora or organizations are necessary to avoid parallel, duplicative work.

When it comes to emergency prevention, preparedness and response in the Arctic, there is generally a lot of goodwill all around, and good initiatives and projects are being developed constantly. But it is a huge task just to identify and follow these initiatives, and there is a very real danger that ongoing work is duplicated by others, or that work taking place in one organization steals attention from other similar projects and initiatives. This is a real risk with the EPPR and the ACGF as well, but through increased dialogue and even personal connections, it has been avoided so far.

It is of the utmost importance that the Arctic states, indigenous peoples, coast guards, and other actors focus on making *active use* of existing fora and organizations. It is the only way to give them the strength and (eventually) financial support to deliver on their missions and to avoid organizational competition.

Now What?

Back to the big storm in 2010 and the missing kayakers. What was the result of the SAR operation with the 14 missing people and the 25 unaccounted for? Thanks to pure luck and to brave efforts by locals and by the tourists themselves, all missing persons were found in a relatively good condition. It was close to a miracle. Now the relevant question is, of course: what has changed since 2010? The frightening answer – Not much! – keeps SAR personnel and authorities awake at night.

Humans are curious by nature and always seek new adventures and opportunities; this will never change. This is why we see intensified traffic and activity in the Arctic region following the increased accessibility caused by changes in the climate and in technology. What we *can* change is the identification, awareness, and management of the risks that attend those adventures and opportunities. And remember the three most important things: co-operation, co-operation, co-operation.

11 Co-ordination of Oil Emergencies in the Russian Arctic

Issues of operational hierarchy and coordination challenges

Svetlana Kuznetsova

Introduction

The risk of oil spills is present in the Arctic (Davidson et al., 2008; Kirby and Law, 2010). Thus, oil spill response (OSR) remains an important issue. The risk of oil pollution results from the growth of oil transport and expected offshore petroleum developments in the Norwegian and Russian Arctic (AMAP, 2007; Arctic Council, 2009; Bambulyak and Frantzen, 2009). In Russia, sea oil transportation and extraction on the shelf have not yet caused significant oil spills. Nonetheless, the preparedness for large oil spills must be enhanced (Zhuravel, 2013).

The Arctic region is an area of special concern regarding oil spills, owing to the extreme weather conditions, vulnerable ecosystems, long distances to land-based port infrastructure, and other specific challenges associated with responding to oil spills in cold environments (Arctic Council, 2009; ENINO, 2014). Any of these conditions increase the consequences of oil spills and at the same time could negate the OSR effectiveness. Additionally, oil spill recovery techniques suffer from reduced functionality under severe Arctic environment (Borch and Andreassen, 2015). As a consequence, understanding the technology and competences for utilizing complex preparedness tools are of critical importance in the High North (Borch and Solesvik, 2013).

The purpose of this chapter is to explore the co-ordination tasks and challenges in oil spill response operations in the context of the Arctic region. We elaborate on how the institutional aspects of the OSR system influences the co-ordination between the companies and agencies involved.

This paper is organized as follows. First, the relevant theories about the Incident Command System (ICS), networks, and mechanisms of inter-organizational co-ordination are presented. Then, the case of the tanker 'Nadezhda' is discussed. In the analysis chapter, we take a closer look into the co-ordination gaps within oil spill response and analyze how the standard management structures of the ICS and the available restructuring mechanisms facilitate the flexible and efficient use of heterogeneous resources. Finally, we draw attention to the implications for industry and future research in this area.

Theory

Oil spill incidents can require a response involving many organizations, including companies and governmental authorities. There could also be parallel activities such as firefighting, protection of the environment, and securing property and infrastructure from damage. This increases the complexity of the OSR activities (Taylor et al., 2008). Thus, oil spill response is a complex and dynamic cross-disciplinary activity that unfolds in a continuously evolving and, at times, highly uncertain environment. The greater the scale of the oil spill, the larger the number of actors involved. The effectiveness of the OSR system depends on the ability of these actors to work together to ensure OSR activities to be conducted (Sydnes, 2011). The interaction within the OSR can call for restructuring for improved co-ordination and to achieve the intended effect (Auf der Heide, 1989; Haas, 1992).

Incident Command System and Networks

Within emergency management, co-ordination between different actors and their incident co-ordinators relies on factors such as agency interdependencies and the established individual and joint management structures and mechanisms for co-ordination and control. Mintzberg defines co-ordination and control as essential mechanisms in organizations (2009). Co-ordination is understood as an emergent process when different interdependent action trajectories are synchronized (Wolbers et al., 2017). Co-ordination is defined as 'both a process – the act of co-ordinating – and a goal: the bringing together of diverse elements into a harmonious relationship in support of common objective' (Seidman and Gilmour, 1986).

The ICS is an emergency management tool that defines specific roles and an organizational hierarchy that facilitates co-ordination and control of diverse resources (Bingley and Karlene, 2001). The basic ICS includes a set of rules and practices to guide actions, common terms for equipment and supplies, a structured chain of command from the specialist at an emergency site to the incident commander, and a designated authority, along with responsibility and task assignments. The span of control is limited to the number of people that one person can effectively control. The division of tasks into sectors is important to ensure effectiveness and safety (Buck et al., 2006). Its adaptable nature and ability to integrate multiple resources quickly into a joint and goal-oriented team are the main advantages of the ICS structure (Lindell et al., 2005). Initially developed for fighting forest fires in California in the 1970s, the ICS has been used in various types of emergency operations resulting from both natural and technological accidents, including oil spill response (Bigley and Roberts, 2001; Lindell et al., 2005).

The ICS is structured to facilitate activities in five major functional areas: command, operations, planning, finance, and logistics. The ICS does not create a universally applicable bureaucratic organization among responders, but is rather a mechanism for inter-organizational co-ordination designed to impose the order on certain dimensions of the chaotic organizational environments of disasters (Buck et al., 2006).

Some authors criticize the ICS for ignoring the importance of inter-organizational relationships, the spontaneous nature of the response, and the potential for conflict between organizations (Drabek and McEntire, 2003; Waugh and Streib, 2006). This criticism is consistent with a co-ordination and communication perspective (Drabek and McEntire, 2003) which points out that any large scale crisis inevitability requires an inter-sectoral and cross-jurisdictional response (Boin and Hart, 2003). The researchers argue that decentralization and flexibility are necessary to deal with the inexactness and turbulence of crises (Moynihan, 2008; Tierney and Trainor, 2004).

Therefore, in the case of inter-organizational response, the ICS is better understood as a means of network governance, established to co-ordinate interdependent partner organizations under urgent conditions, but not as a hierarchical structure (Provan and Kenis, 2008; Provan and Milward, 1995). Network theorists propose that the ICS is an example of a highly centralized model of network governance (Moynihan, 2009). The incident command does not exist independent of the network; indeed, its purpose is network governance. To meet the critique, Bigley and Roberts (2001) highlight structuring mechanisms that formalize this process in high complexity environments. These mechanisms are structure elaboration, role switching, authority migrating, and system resetting. The structure elaboration process is an initial one, because management is organized on an emergency site. Role switching is the process of assignment and reassignment of personnel to different positions according to the functional requirements of the emergency operation. The authority migration happens when critical expertise or capacity in a specific emergency area is moved from the official hierarchy authority to another one, when needed. System resetting makes the organization look through the

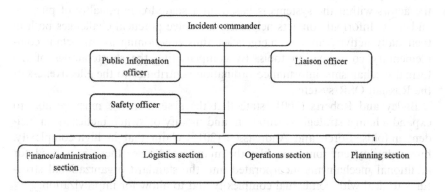

Figure 11.1 ICS basic organization chart (US Federal Emergency Management Agency)

structure, competence, and routines, in the light of the changes within the operation to match them to working conditions (Andreassen et al., 2018).

Formal and Informal Co-ordination

The organization theory literature emphasizes that the priority role of co-ordination in inter-organizational collaboration and networks results from its continuous synchronization of tasks and the contribution of organizations involved. It constitutes a relational process through systematic and reliable communication which strengthens informal relations beyond rigid administrative structures to achieve better integration, trust, and commitment (Kozuch and Sienkiewicz-Małyjurek, 2016).

The Russian OSR system includes a large number of organizations that could be involved in an emergency. The OSR system is multi-organizational, and the tasks and functions need to be integrated. Sound organization and effective inter-organizational co-ordination among the key actors become both an advantage and a challenge for successful oil spill emergency response. According to Moynihan (2009), the most evident complication that a network setting brings for the ICS is the number and diversity of organizations involved. This intensifies the heterogeneity of backgrounds and interests of network members.

Sydnes (2011) illuminates how a simultaneous interplay of formal and informal mechanisms helps to avoid the current prescriptive approach to OSR planning in Russia. This fact manifests itself in the requirements enforced by regulatory authorities for performance and monitoring that the operators comply with (ENINO, 2014). These legislative requirements define the formal co-ordination while formalizing the structure, roles, and functions of the interacting organizations. It also establishes the operative co-ordination procedures and patterns of interaction in an emergency (Sydnes, 2011).

Informal co-ordination is common among actors well known to each other and compensates for the lack of formal procedures. Interdependence among the actors within the systems is based on a shared commonality of purpose and trust. Informal contacts help actors to solve practical challenges both in their daily activity and in emergencies, thus functioning as important complementary co-ordination tools. By compensating the shortcomings of the formal mechanisms, informal co-ordination contributes to the effectiveness of the Russian OSR system.

Bigley and Roberts (2001) state that the system in use must be able to expand, change strategic orientation, and modify or switch tactics as an incident unfolds. Borch and Andreassen (2015) claim that in high-complexity, high-volatility environments like the maritime Arctic, there is a need for additional mechanisms incorporated into the standard organizational structures to deal with contextual complexity and to allow for improvization.

In Russia, the theory on the ICS, networks, and the structural mechanisms of the co-ordination is not widely used within the oil spill response agencies.

In the following section we analyze the co-ordination of OSR considering the organizational challenges resulting from the complexity and dynamism of the Arctic.

Methodology

In this study, we build upon an illustrative case study of an incident related to major oils spill in cold conditions. We provide data from the grounding of the tanker *Nadezhda* south-west of Sakhalin in 2015. The data for analysis includes reports and articles, media releases, contingency plans, and a detailed description of the accident (WWF, 2017). The data is also verified with interviews with key personnel at the Maritime Rescue Co-ordination Centres (MRCC) and Subcentres.

Data and Analysis

The Nadezhda' *Accident South-west of Sakhalin*

On the morning of 28 November 2015 a powerful storm blew the tanker *Nadezhda* onto a reef off the northern coast of Nevelsk in Sakhalin oblast in the Far East of Russia, located some 150 metres offshore outside of the port's water area. The *Nadezhda*, which was built in 1986, was carrying 360 tons of heavy fuel oil and 426 tons of diesel fuel. The two centre tanks contained fuel oil, and the rear tanks with diesel oil were damaged. None of the eight crew members on board, including the master, was injured.

Three private oil response companies were called on to respond to the oil spill within the shoreline. One of them had an agreement with the harbour authorities to provide OSR but did not possess the necessary resources. The same urgent order was issued to the local marine rescue service (Morspassluzhba) with an OSR-equipped tugboat, *Rubin*.

On the same day, a region-level emergency regime was established within the port area of Nevelsk by the head of the Sakhalin region emergency commission/the deputy head of the regional government. The Nevelsk Harbour Master was to take the OSR measures prescribed by 'The Oil Spill Prevention and Response Plan'; and the Sakhalin oblast's Minister of Finance to ensure funding. The municipal public water supply company was responsible for preparing tanks and equipment needed to pump out oil products from the damaged tanker. Rospotrebnadzor (the federal consumer rights and human wellbeing body) had to control environmental conditions within the coast. The municipal dispatch service was ordered to keep the public informed.

The bodies that eventually took responsibility for the co-ordination of the OSR operations on board the tanker were the Marine Port Authority and the Nevelsk Harbour Master. On the shoreline, the Ministry of the Russian Federation for Civil Defense, Emergencies and Elimination of Consequences of

Natural Disasters' (EMERCOM) administration for Sakhalin was ordered to ensure the preparedness of all the resources, to collect and integrate the emergency-related information, and to keep the public informed.

Approximately one day after the accident, the decision was taken to conduct ship-to-shore transfer of the oil products. The operations were interrupted by stormy weather several times. On the night of 4 December, another cyclone struck the whole Sakhalin peninsula, causing more damage and sending the tanker to the bottom of the sea after it lost buoyancy.

The diesel fuel transfer was completed on the 11th day after the accident. Out of the 560 tons of the on board diesel fuel, only 345 tons were pumped out. Given all of the information available, the volume of the oil spilt can be estimated at a minimum of 250 to 300 tons and approximately 100 tons of diesel fuel.

Oil coated a 20-km stretch of the shore, with the sticky sludge extending up to 4 metres on land from the water line. This resulted in the deaths of animals and birds. Among the teams involved in the clean-up operations on the coast were private companies and the Sakhalin branch of Morspassluzhba. The teams that dealt with the affected animals were volunteers, with support from Nevelsk's municipal administration (WWF, 2017).

The Russian Oil Spill Response System

The Russian OSR system is based on a structure conducted at multiple levels by the federal executive authorities, the administrations of the Russian Federation's sub-units (including local administrations) and oil companies (Government of the Russian Federation, 2000; Government of the Russian Federation, 2002).

Russian legislation classifies an oil spill as a 'state of emergency' (Government of the Russian Federation, 2000). All questions related to emergencies in Russia, including OSR activity, are the remit of federal authorities and are organized and performed in the framework of the Integrated National Emergency Prevention and Response System of the Russian Federation.

The OSR system is divided into sea and land sectors that function under the auspices of two different ministries – EMERCOM for the shore and land, and the Ministry of Transport for the seaside and maritime sector. Both subsystems are intended to work independently, according to their plans (Government of the Russian Federation, 2002).

The Ministry of Transport is responsible for OSR at sea (Government of the Russian Federation, 2003). This authority is composed of several federal agencies. Subordinate agencies to the Ministry of Transport are the Federal Agency of Maritime and River Transport (Rosmorrechflot) and the Morspassluzhba. Rosmorrechflot carries out the general management of the OSR system at sea. At the same time, Morspassluzhba controls the daily operational activity of the system. Its rescue divisions in the region respond to oil spills at sea (Ministry of Transport of the Russian Federation, 2009).

Oil spill operations at sea are co-ordinated by the State Maritime Rescue Co-ordination Centre reporting to Rosmorrechflot on the federal level, by the MRCCs and Sub-centres on the regional level. On the local level, OSR operations are co-ordinated by dispatcher centres of maritime transport organizations, ports, shipping companies, and other organizations engaging in petroleum exploration, production, processing, and transportation. The ICS is not widely adopted by the OSR emergency organizations in Russia. Mostly, the interaction charts from all the services involved are used. An example from the Murmansk MRCC is shown below.

Another main actor is the Ministry of Natural Resources and Ecology, which is responsible for policymaking, control, and supervision related to the study, use, reproduction, and protection of natural resources and the environment. Two federal services perform control and supervision: the Federal Supervisory Natural Resources Management Service (Rosprirodnadzor) and the Federal Service for Ecological, Technological, and Nuclear Surveillance (Rostehnadzor). Rosprirodnadzor is subordinate to the Ministry of Natural Resources and Ecology, while Rostehnadzor reports directly to the government. Together with EMERCOM, the Ministry of Natural Resources and Ecology classifies oil spills and thereby decides how much the polluting party will be fined (Ministry of Natural Resources of the Russian Federation, 2003). Obligatory guarantees of sufficient funds to respond and compensate the environmental damage were set recently in the legislation.

Organizations that engage in petroleum exploration, production, processing, and transportation are obliged to ensure oil spill response either through their own dedicated divisions or external certified contractors. They are also obliged to have the standby funds and material resources necessary to localize and liquidate oil spills available in exploration, production,

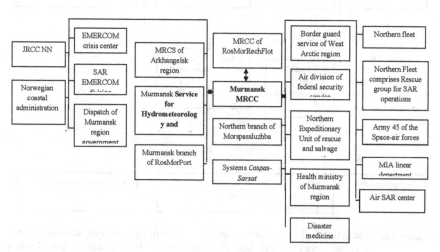

Figure 11.2 SAR interaction in the Murmansk responsibility area

processing, and transportation zones (Government of the Russian Federation, 2002; Government of the Russian Federation, 2004).

Oil spills are classified by the Russian legislation in terms of their potential severity. Based on the volume of oil or oil products spilled at sea, emergencies can be classified as:

Managed on local level: oil or oil product volumes under 500 tons
Managed on the regional level: 500–5,000 tons of oil or oil products
Managed on the federal level: more than 5,000 tons of oil or oil products

Government of Russia, 2000

Local oil spills are detected and eliminated by the organizations engaging in petroleum exploration, production, processing, and transportation. If oil spill volume exceeds the highest-rated volume determined in the contingency plan of the oil companies, and the organization is not able to localize and remove the oil spill, the regional OSR plan comes into action. The emergency divisions of Morspassluzhba have standing response resources on the regional level in case of oil spills. A similar procedure is applied if an oil spill extends up to the federal level.

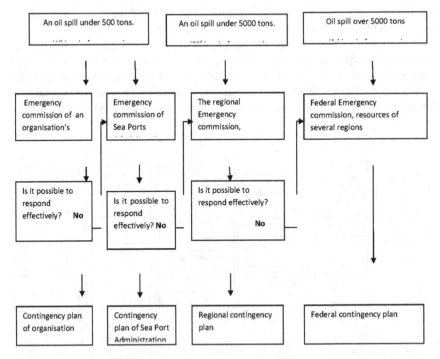

Figure 11.3 Patterns of shifting responsibilities according to oil spill volume – contingency plan of the Sea Port of Naryan-Mar (Nenets Autonomous Okrug)

As oil spills are classified depending on the volume of oil spilt, contingency plans are worked out for combating spills of different levels (federal, regional, local) and are enacted depending on the category of the oil spill. *Inter alia*, the oil spill contingency plans provide an algorithm of actions and co-ordination to be taken during an emergency response operation and thus are meant to facilitate preparedness.

A key element of the Russian OSR system is that its organizational structure changes in the emergency mode. In the steady-state mode, it functions as a contingency bureaucracy ready to provide its services if an emergency occurs (Mintzberg, 1983). In emergencies, an additional institutional body, the Emergency Commission, plays a role. It is a co-ordinating body to be assembled in the event of an emergency, including an oil spill. The Emergency Commission is presented on different levels and headed by the regional government or federal ministry. The main task of this body is to mobilize, organize, and bring together all available resources and organizations needed for effective OSR. It unifies the sea and land sectors and also includes representatives of the technostructure and support staff when needed.

In other words, in the emergency mode, the organizational structure is simplified and becomes more linear by appearing as the Emergency Commission. Now the agency is pulled in the direction of the strategic apex (because the power of the federal authorities in the Russian system is strong, and most regional-level actions are subject to their approval). The operating core is where the strategic apex integrates the components of technological resources and staff resources (Sydnes, 2011).

A structured chain of command from the specialist at an emergency site to the incident commander is not clear. The system reflects the network nature of the inter-organizational co-ordination of OSR in Russia. The Commission is a platform that integrates all structural components of the system. At the same time, the members retain a significant element of autonomy. Thus, the Emergency Commission head might not necessarily serve as an incident commander.

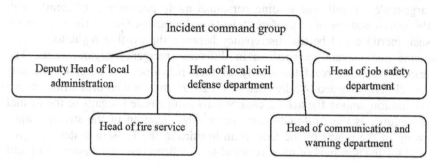

Figure 11.4 A version of an Emergency Commission

Discussion

In this section, we analyze the main co-ordination challenges, the assignment of tasks to different actors and the communication aspects in the Nadezhda incident, on the different stages of the operation.

The grounding stage of the vessel shows that pre-planned responsibilities can be different when facing a real accident. Because of the quickly changing working conditions owing to the stormy weather that influenced on the action plan, co-ordinators had to rearrange their procedures when the responsible body, a private company, Island General Services, was not able to provide efficient response. The number of spilt oil products was under the responsibility of the owner and this private oil response company, which had an agreement with the port of Nevelsk to provide first-level OSR.

As the ship's crew and responsible company did not manage to maintain command and were not able to co-ordinate efforts at the emergency site, the municipality, regional contingency plans, and the contingency plan of the port of Nevelsk were activated. At the OSR stage, the mechanism for structure elaboration was referred according to demanding operational circumstances when the Emergency Commission was convened.

The Emergency Commission head struggled to incorporate and direct the extended network to distribute the tasks and functions to respond to oil product spill and further clean-up and recovery. The range of tasks led to the creation of some task-specific networks within the broader response network, dealing with goals such as delivering materials; financial support; public safety; and information-sharing. There were basic problems in co-ordination both within and across these networks because participants brought the perspective of their home organization. Because of organizational differences, the organizational plans and procedures of one organization clashed with the perspectives of other network members. Each organization highlighted their autonomy and their primary institutional identity within their home organization rather than the temporary Joint Incident Command.

The existing system of maritime assistance in Russia is distinguished by network governance, established to co-ordinate interdependent partner organizations under urgent conditions. However, in new types of accidents like large-scale oil spills, the leading personnel might not interact efficiently with the representatives of other departments and agencies. One of the reasons for such inertia could be the discrepancy between the existing regulatory documents and current instructions in different emergency organizations. The responders either react too slowly, or their help does not cover all the personnel that is needed (Andreassen et al., 2019). Informal co-ordination, which is common among the actors, cannot fully compensate for gaps in the formal procedures as the *Nadezhda* case shows. The direction of the strategic apex owing to the power of the federal authorities in the Russian system and the need for the approval of most regional-level actions can overshadow trust and shared purpose.

The co-ordination gaps challenge the implementation of the effective co-ordination mechanisms such as authority migration, role-switching, and system-resetting. The lack of organizational flexibility could hinder the movement of critical expertise or capacity from one official hierarchy of authority to another and restrain the assignment of personnel to different positions.

These gaps are clearly reflected in the information flow in the *Nadezhda* case when several organizations – the harbour office, EMERCOM, municipal press service, etc. – functioned as spokesmen simultaneously and sent information to the public not co-ordinating with each other. Hence, because of breakdowns in inter-organizational information-sharing among different actors, the data about the oil spill itself and the oil spill response efforts were contradictory.

In the case of the *Nadezhda* accident, we can highlight several challenges that caused inefficient co-ordination. The most important challenges were the limited and unprepared resources, the lack of standard OSR procedures, and the gaps in networks interaction. The established oil spill response organizations does not include the integration mechanisms found in the ICS scheme. Taking into account network diversity and the differences in backgrounds of the responders, it is critical to have a common language and set of management concepts to bridge these differences.

Arctic Context

The *Nadezhda* case happened in cold conditions but not in the Arctic. The oil spill occurred near the coastline and in a well-developed area with many resources available. However, it caused essential environmental damage. The Arctic is characterized by extreme natural and climatic conditions, including drift ice in the Arctic Ocean; diverse development of the regional economy; low population density (1–2 per 10 km^2); remoteness from major industrial hubs; and the vulnerability of natural ecosystems to human-made disasters. All of these factors make the emergency response much more challenging. The current state of the Arctic zone development in the Russian Federation is defined by limitations in equipment and technologies and insufficient development of navigation, hydrographic, and hydrometeorological support, as stated in the development strategy of the Arctic zone.

It should be mentioned that the consequences of accidents such as loss of lives, environmental damage, and economic loss could be more severe in the Arctic because of great public attention put on all Arctic activities, low public tolerance for accidents, and the potential loss of reputation for all parties involved (Arctic Monitoring and Assessment Programme, 2007).

To summarize, regional, climatic, and situational peculiarities could challenge OSR in the Arctic. A lack of experience in managing OSR in extreme conditions and limited resource capacities could hinder an efficient oil spill response. Available facilities and equipment should be optimized for the northern climate conditions. There is a need to focus on well-developed management systems, such as the ICS and the necessary co-ordination mechanisms to provide efficient OSR.

Conclusion

Large-scale maritime oil spill incidents in the Arctic could result in an overload in the normal emergency response system. Limited resource availability, resource-consuming mobilization time, and the lack of experience in this kind of incidents in the High Arctic context could put a heavy strain on the management levels.

The network interaction perspective reflected in the Emergency Commission concept could be subject to ambiguity and disagreement. The involved co-ordinators should monitor the operational environment and share information that would help the overall situational awareness, despite limitations in information exchange capacities. Flexibility in the decision-making process is vital at all management levels, including finding new resources and solutions, as well as adapting standard operating procedures to the prevailing environment and using local know-how and resources.

In this chapter, we have highlighted the need for introducing inter-organizational restructuring mechanisms and the principles of the ICS, allowing for flexible on-scene co-ordination of emergency response to meet the challenges of the Arctic environment.

Authority delegation, assembling and reassembling task forces, role-switching, and system-resetting are critical prerequisites for an efficient and flexible OSR co-ordination in the Arctic. However, these requirements could call for further regulatory changes and competence-building in the maritime Russian OSR system.

Implications for Industry

The discussion in this chapter shows the strain on the emergency management system of the Arctic region in a significant OSR crisis. Industries have launched particular standards for polar waters like the oil and gas industry's 'Incident management system for the oil and gas industry' (IPIECA, 2014). The Polar Code provides additional demands for vessels operating in the polar regions. However, current regulations could be seen as minimum standards for the High North. Thus, the OSR authorities should emphasize further development of roles, routines, and procedures, including the development of the ICS. In the case of a large-scale incident, the incident commander can designate one or more command staff positions to perform various managerial roles. Such positions can be related to public information, safety, liaison, legal, human resources issues, or finance to manage the insurance issues that are often an important part of oil spill response. Some additional units for the operations section can be recognized as offshore operations and onshore operations ones. An environmental unit can be created in recognition of the efforts required in managing the multitude of environmental issues associated with oil spills. A central function of this unit can involve the collection and evaluation of operational information about the incident, including the current and forecast situation and the status of assigned resources. Furthermore, it is of great importance to increase the degree of competence-building both in private companies and between them and government authorities in co-ordinating OSR.

The Implication for Further Research

In the chapter, we have built upon one illustrative case. There is a need for quantitative studies demonstrating the contextual elements and their influence on the co-ordination issues and structuring mechanisms. In particular, one should elaborate on the resource re-configuration with a mix of capacities from various sources, including the resources from neighbouring countries.

There have been efforts to improve both capacities and competences in the Arctic regions in the OSR system, but further gap analyses on response time and technologies should be provided. The annual Exercise Barents between Norway and Russia gives valuable information, as well as trust and competence exchange. Also, there is a need for comparative studies looking into the planning, operational challenges, and competence needed for all the maritime areas in the Arctic.

References

Andreassen N., Borch O. J., Ikonen E. (2018). Managerial Roles and Structuring Mechanisms within Arctic Maritime Emergency Response. *Arctic Yearbook 2018*.

Andreassen, N., Borch, O. J., and Ikonen, E. S. (eds.) (2019). Organizing emergency response in the European Arctic: a comparative study of Norway, Russia, Iceland and Greenland MARPART Project Report 5. Bodø: Nord Universitet 2019, R&D report N46. http://hdl.handle.net/11250/2611539.

Arctic Council. (2009). *Arctic Marine Shipping Assessment 2009 Report*. Akureyri.

Arctic Monitoring and Assessment Programme. (2007). Arctic Oil and Gas 2007. Oslo.

Auf der Heide, E. (1989). *Disaster Response: Principles of Preparation and Coordination*. 1989. St Louis, MO: C.V. Mosby Company.

Bambulyak A., Frantzen B. (2009). Tromsø: Akvaplan-Niva.

Bigley, G. A., Roberts, K. H. (2001). The incident command system: high reliability organizing for complex and volatile environments. *Academy of Management Journal*, 44(6): 1281–1299.

Boin A., Hart P. (2003). Public leadership in times of crisis: mission impossible? *Public Administration Review*.

Borch O. J., Andreassen N. (2015). Joint-Task Force Management in Cross-Border Emergency Response. Managerial Roles and Structuring Mechanisms in High Complexity High Volatility Environments.

Borch, O. J., Solesvik, M. (2013). Collaborative Design of Advanced Vessel Technology for Offshore Operations in Arctic Waters. *Computer Science*, 8098(1).

Buck, D. A., Trainor, J.E., and Aguirre, B.E. (2006). A Critical Evaluation of the Incident Command System and NIMS. *Journal of Homeland Security and Emergency Management*, Vol. 3, Issue 3.

Davidson, W. F., Lee, K., and Cogswell, A. (2008). Oil spill response: a global perspective.

International Maritime Organization. (2005). Manual on oil pollution (section IV: combating oil spills). London.

Donald P., Moynihan D. (2009). The Network Governance of Crisis Response: Case Studies of Incident Command Systems. *Journal of Public Administration Research and Theory.*

Drabek T. E., McEntire, D. A. (2003). Emergent phenomena and the sociology of disaster: lessons, trends and opportunities from the research literature. *Disaster Prevention and Management.*

ENINO. (2014). Russian – Norwegian Oil & Gas industry cooperation in the High North, Environmental protection, monitoring systems and oil spill contingency.

Gilmour R., Seidman H. (1986). *Politics, Position, and Power: from the Positive to the Regulatory State.* New York, Oxford University Press.

Government of the Russian Federation. (2000). Resolution 613, 'About Urgent Measures on Prevention and Combating Oil Products and Oil spills'. 21 August.

Government of the Russian Federation. (2002). Resolution 240, 'On procedure on Implementation of Measures to Prevent and Response to Oil and Oil Product Spills at the Territory of the Russian Federation', 14 April.

Government of the Russian Federation. (2003). Resolution 794, 'On the Integrated National Emergency Prevention and Response System of the Russian Federation'. December.

Government of the Russian Federation. (2014). Resolution 1189 'On oil and oil products spill prevention and liquidation organization on the RF continental shelf, on inland sea waters, in the territorial sea and adjoining zone'. 14 November.

Haas , P.M. (1992). *Introduction: Epistemic Communities and International Policy Coordination.* Cambridge: Cambridge University Press.

IPIECA. (2014). Incident management system for the oil and gas industry (manual).

Kirby, M. F. and Law, R. J. (2010). Accidental spills at sea – risk, impact, mitigation and the need for co-ordinated post-incident monitoring. *Marine Pollution Bulletin*, 60(6),: 797–803.

Kozuch B., Sienkiewicz-Małyjurek, K. (2016). Inter-Organisational Coordination for Sustainable Local Governance: Public Safety Management in Poland. *Sustainability*, 8(2): 123–139.

Lindell M. K., Perry R. W., and Prater, C. S. (2005). Organizing Response to Disasters with the Incident Command System/Incident Management System (ICS/IMS). International Workshop on Emergency Response and Rescue.

Mintzberg, H. (2009). *Managing.* Williston, VT: BerrettKoehler Publishers.

Moynihan, D. (2008). Learning under uncertainty: Networks in crisis management. *Public Administration Review*, 68(2): 350–365.

Nenets Autonomous Okrug. (2011). Contingency plan of the Sea Port of Naryan-Mar.

Ministry of Natural Resources of the Russian Federation. (2003). Order 156, . 'On Approval of Guidance to Determine the Lower Level of an Oil Spill and Oil Products to the Assignment of Accidental Spills Emergency'. 3 March.

Ministry of Transport of the Russian Federation. (2009). Order 153, 'On Approval of Policy Directive on the Functional Subsystem of Prevention and Response Organisation of an Oil Spill at Sea'. 6 April.

Provan K. G., Kenis P. (2008). Modes of network governance: structure, management, and effectiveness. *Journal of Public Administration Research.*

Provan K. G., Milward H. B. (1995). A Preliminary Theory of Interorganizational Network Effectiveness: A Comparative Study of Four Community Mental Health. *Systems Administrative Science Quarterly.*

Sydnes M. (2011). Oil spill emergency response in the Barents Sea: Issues of interorganizational coordination. Thesis, University of Tromsø.

Taylor E., Steen A., Meza M., and Couzigou F. (2008). Assessment of oil spill response capabilities: a proposed international guide for oil spill response planning and readiness assessments. The 20th Triennial International Oil Spill Conference (IOSC) on Prevention, Preparedness, Response and Restoration, Savannah, GA, 4–8 May 2008.

Tierney K., Trainor J. (2004). *Networks and resilience in the World Trade Center disaster. Research Progress and Accomplishments 2003–2004*: p. 157–173.

Waugh, Jr W. L., Streib, G. (2006). Collaboration and leadership for effective emergency management (Public administration review).

WWF. (2017). What Nadezhda can teach us. Retrieved from: https://wwf.ru/en/what-we-do/green-economy/what-will-nadezhda-teach-us/.

Wolbers, J., Boersma, K., and Groenewegen, P. (2017). Introducing a Fragmentation Perspektive on Coordination in Crisis Management. *Organization Studies*: 1–26.

Zhuravel, V. (2013). Risk of emergency oil spills at sea and contingency plans to prevent and respond. *Offshore Russia*, September.

12 Co-ordination of Emergency Response Systems in High-Complexity Environments: Structuring Mechanisms and Managerial Roles

Natalia Andreassen and Odd Jarl Borch

Introduction

The co-ordination of joint operations between many independent organizations is a challenging managerial task for the incident commanders in charge. Emergency response operations often include a broad range of agencies which collaborate closely in operations that can be characterized by high complexity and volatility. Large-scale emergency responses can demand resources beyond that required by the average type of accident. The lead agencies might have to mobilize a broad range of air, land, and sea-based personnel and equipment. Support from other countries could be needed. The external resources can belong to different agencies with different expertise, role structures, plans, and standard operating procedures (Moynihan, 2009; Beck et al., 2014; Wolbers et al., 2018).

The participants within emergency response such as search and rescue (SAR) operations can include rescue co-ordination centres, fire and rescue services, police, military forces, paramedics, and private rescue organizations, as well as volunteers. Transparent management and efficient co-ordination between several agencies are challenging tasks for the commanders in charge (Bigley and Roberts, 2001; Wolbers et al., 2018).

In the Arctic regions, resources might be scarce, and the distances between the scene of an accident and the resource bases are long. Equipment functionality could be reduced because of the cold climate. Harsh weather conditions can cause unpredictability and a very challenging operational context, for example related to logistics (Marchenko et al., 2018). This context can create a high degree of complexity with a lot of factors to be considered by the involved agencies. Task complexity is defined as the number of components and the ties between them that can provide alternative routes towards a goal (Campbell, 1988; Hærem, 2015).

The starting point for this study are the interlinking chains between a multi-agency task force created from a complex web of various institutions and the organizational structure and managerial roles of the operation (Borch and Andreassen, 2015; Andreassen et al., 2018a,c).

In this chapter we explore how specific structuring mechanisms have to be prepared within the lead organizations to facilitate the adaption of management to the context complexity. We claim that task complexity within emergency response can call for both deviation from and addendums to the standard operating procedures. Among the lead emergency agencies in the Arctic countries we find a range of organizational structures and formalized procedures, especially between the land and sea-based response agencies (Borch and Andreassen, 2015: Andreassen, Borch, and Ikonen, 2019). Several studies have emphasized the challenges of structuring emergency response and especially the limitations of standardized command systems within large-scale emergency response (Bigley and Roberts, 2009; Moynihan, 2011; Turoff et al., 2011). In this study we emphasize the inter-organizational structuring mechanisms including task division, hierarchical authority, and the formalization including plans, rules, regulations, and procedures. Secondly, we illuminate how the structuring mechanisms available can facilitate the necessary changes in the management composition to prepare for the unknown.

In the next sections, we reflect on the platform for studying structuring mechanisms and the managerial roles of incident commanders. In the data and analysis section, we take a closer look at the operational tasks and management responsibilities of the organizations involved in challenging operations. Based on the two maritime SAR cases of the *MV Northguider* and *MV Viking Sky*, we reflect on the emerging structural mechanisms that can allow for the system's adaptation and improvization to create a context-adapted emergency response task force.

Theoretical Framework

Emergency Response Task Complexity and Value Chains

In large-scale emergencies, a broad range of organizations have to be mobilized and united to form an effective, fast-responding emergency response value chain. (Bigley and Roberts, 2001; Moynihan, 2009). Inter-organizational collaboration can be seen as co-ordinated interaction between different organizations towards a common goal. The organizations are acting towards solving a common set of problems with shared resources, knowledge, rules, and structures (Beck and Plowman, 2014; Phillips et al., 2000). In emergency response, the joint command of a composite task force has to make decisions under time pressure, and the teams have to be effective under often difficult and sometimes dangerous conditions (Flin et al., 1996; Crichton et al., 2005). This context calls for an efficient organization and superior management at all levels and in all functions. As a starting point, we may find that the involved organizations have their own set of goals and values, structure, operational mode, and operation procedures. Even though some structures could be based on the same set of professional standards, rules, and regulations, a certain diversity in practice could call for adaptations. In cases where a situation is escalating, mobilization and inclusion of capacities that have not necessarily worked together before could be necessary. Thus, task reorientation and flexible couplings between units in tailor-made value chains could be called for (Beck and Plowman, 2014).

Value chains within emergency preparedness refer to the set of activities that contributes to the final goal of saving life and health, protecting the environment, and material, economic, and social values. The value chain perspective focuses on the pattern of interconnectedness of several units and their contributions in providing the final value for the operation. The inter-action between the agencies in an emergency response value chain could vary, according to task complexity. The emergency response agencies could be integrated into sequential value chains or in close-knit value networks (Moynihan, 2009). They can work in parallel, performing their tasks with limited interplay. They can work sequentially where one agency takes over from the other, passing information about what is done and what can facilitate the transaction for the next tasks and responsibilities. Equally, the agencies can collaborate closely with high inter-dependency in value networks, where the participating organizations are intertwined in reciprocal relations. The complexity of the operation could affect the value chain configuration.

Task complexity can stem from this need for a high degree of integration within the value-network. Complexity is increased as a result of the number of actors, the diversity of actors, and ties between them, causing uncertainty about the cause–effect relationship, circular causality, and the dramatic effects from small changes (Erdi, 2008; Weick and Sutcliffe, 2011.) As Hærem et al. (2015) claim, "the complexity of a collective task is not merely the sum of the complexity of the constituent tasks, because interdependence between multiple actors can have an exponential effect on task complexity". They regard complexity as a function of the network of events involved in a task. In large-scale emergency response, this number of actors, events, and the ties between them could be significant from the start and increase during the action.

The combination of high complexity and volatility could represent a challenge to the entire value chain because each of the actors involved could have to take steps to re-position during the action. In doing this, they have to adapt their value chains and re-configure their response resources. This could cause ripple effects in the interlinked value chains. Thus, for the lead partner, it is necessary to 1) look into each part of the value chain at the functions, their processes and how they are interlinked; and 2) have the overview over the whole system and its totality in efficiently providing values. In addition, the emergency response system could be influenced by and dependent on more loose couplings, among others to the range of stakeholders in an operation (Borch and Batalden, 2014). The leaders might have to deal with the media, next of kin, owners, interest groups, and politicians as parallel processes in not only local but national and international arenas, including the homelands of persons and units involved (Boin and t'Hart, 2003; Crichton et al., 2005).

If an upscaling of the response is needed with more tasks and resources involved, several people at different management levels could be involved in co-ordination over time. The awareness of how the collaborating units might be interconnected not only internally but also to external parties is important to understand the role change and necessary adjustments in responsibilities, tasks, collaborative skills and management capacities (Boin and Hart, 2003; Tierney and Trainor, 2004; Moynihan, 2009). The increased complexity of the value chain could call for improvization in management with a certain degree of freedom from pre-established procedures.

Co-ordination and Managerial Roles

Emergency response co-ordination efforts have to be performed both vertically between the different levels of management and horizontally between the different teams and organizations involved. The co-ordination process has to integrate a set of interdependent tasks into a unified arrangement towards the efficient achievement of a common goal (Okhuysen and Bechky, 2009, Wolbers et al., 2018). In a large-scale emergency response, a broad range of tasks has to be performed, often simultaneously, with an escalating number of resources included. With more actors involved and fast reactions needed, the action trajectories of otherwise independent organizations have to be guided by the management levels and synchronized into united performance. This process of coupling new actors with the present actors often has to take place 'on the run' after the response has started (Bigley and Roberts, 2001; Crichton et al.; 2005; Wolbers et al., 2017). To achieve this coupling and decoupling, a smooth interplay between the management levels is critical (Borch and Andreassen, 2015).

The incident commanders in charge of the emergency response system at different levels face several challenges. A broad range of managerial roles has to be performed. A managerial role can be regarded as a type of responsibility and action assigned to a position. Some of the roles can be predefined and performed according to established organizational structures and procedures, while others have to be adjusted or invented during the process (Wolbers et al., 2018). The literature emphasizes managerial roles like situation awareness and communication (Crichton et al., 2005), decision-making under uncertainty (Klein et al., 1986), resource allocation and task delegation (Wolbers et al., 2018), teamwork and trust building (Crichton et al., 2005; Moynihan, 2009), and system improvization (Bigley and Roberts, 2001). Mintzberg (1973, 2003, 2009) categorized managerial roles within an organization into three main groups: interpersonal, decisional, and informational. The formal authority is the starting point for the roles and defines the position of the persons involved. By distinguishing the roles, it is possible to explain the varying nature of tasks inside and outside the units of an organization for the pursuit of the given objective.

Interpersonal roles include serving as the figurehead whose role is to motivate and inspire inside the organization and represent the organization externally, for example to stakeholders such as media, next of kin, etc. Among the interpersonal roles, we find leadership duties towards the teams, not least empowering the subordinates; this also includes putting the person with the right qualifications in charge of the right team (Crichton et al., 2005). The liaison role, which establishes contacts to other relevant organizations, is also crucial in creating a common sense-making process, negotiating resource allocation, and providing expert advice (Bigley and Roberts, 2001; Wolbers et al., 2018). For example, linking up to other levels of responsibility such as the national government to achieve policy-level co-ordination and long-term resource allocation is important.

The Information role is critical in emergency response systems. This role includes the monitoring function which scans the environment and receives different types of information, and dissemination throughout the system (Mintzberg, 2003). In emergency response, the co-ordinators continuously work with the incoming information, creating adequate and shared situational awareness, looking into an alternative combination of resources, and allocating tasks. The informational needs can vary within different operating environments (Paton and Flin, 1999).

Wolbers et al. (2018) emphasize the parallel sense-making processes that take place, resulting in a multiplicity of interpretations. The negotiation around the relevance of different views and perspective that should dominate the decision process might not be seen as dysfunctional. Opening up for negotiations between contrasting views, perspectives, and data interpretation can instead provide a better understanding of the complexity and the differences between the detached partners.

This information processing into a common situational awareness is imperative for the decision-making role (Turoff et al., 2011). *Decision-making* includes both fast choices between alternatives as well as entrepreneurial processes to initiate new action patterns based on the context and the information received. Within the decision-making role, linking up to and choosing between the most important processes of decision-making, finding new avenues, handling disturbance, and responding fast to imminent or unpredicted problems is the challenging part.

Within emergency management, decision-making is a critical input to all the other management roles and functions (Cosgrave, 1996; Paton and Flin, 1999). Decisions have to be made at short notice and as fast as possible to reduce the adverse effects of an emergency.

Klein et al. (1986) presented the recognition-primed model, claiming that in pressing situations, decision-makers can skip the comparison of choices. The approach would be to rely on their experience from similar actions, performing pattern recognition and finding the primary causal factors that define the incident. The decision-maker can perform a mental simulation where the experience-based alternatives are tested on the situation and adapted to the case until a satisfying solution is found (Simon, 1957; Klein, 2008). Through recognition-primed decision-making, the leader could react fast and limit the damage potential and the escalation of the situation. A fast-track decision like this can link up to the most adequate standard operating procedures or create the platform for new solutions. A most demanding decision would be whether or not to deviate from a given management structure and operating procedures (Bigley and Roberts, 2001; Moynihan, 2009; Jensen and Thompson, 2016). Creating new structures, action plans, and procedures and reallocating tasks and teams may be needed (Wolbers et al., 2018). Within emergency management, a specific set of managerial tasks related to co-ordination are important within different incident command systems (ICS). The tasks can become more difficult and co-ordination challenged by highly complex and volatile environments (Hossain and Uddin, 2012; Bigley and Roberts, 2001). For example, the persons holding a specific role or whole teams could be absent or taken out. Several studies have questioned the rigidity of the standardized systems for management control including the ICS implemented by several national agencies and the international management system for SAR related to sea and air incidents (The IAMSAR Manual) (Jensen and Thompson, 2016; Borch and Andreassen, 2015). In high-complexity and high-volatility environments, the organizations should be primed for adjustments in the structure and value chain tasks to increase flexibility in the managerial roles. The lead organization should prepare for structuring mechanisms that contribute to system flexibility.

Standard Operating Procedures and Structures versus Flexibility

While managerial roles refer to a set of certain types of actions, the structuring mechanisms refer to the organizational tools that adapt the organization to a new setting or problem. (Bigley and Roberts, 2001; Buck et al., 2006; Bharosa et al., 2010). Standardized systems such as the ICS have been established to meet new challenges. A high-complexity setting could cause the need to upscale, trigger conflicts, and confusion regarding authority and responsibility. Lack of adequate resources or fast incorporation of auxiliary forces such as volunteers, can lead to communication fall-out and isolating some organizations or teams (Jensen and Thompson, 2016; Andreassen, Borch, and Ikonen, 2018, 2019).

Bigley and Roberts (2001), in their study of ICS, call for more adaptable organizations in emergency response. Jensen and Thompson (2016) take this review a step further. Even though many emergency response agencies have included a standardized command, co-ordination and control system, research shows that there is variety in the implementation of the otherwise standardized systems (Jensen and Thompson, 2016):

> The research suggests that, despite its allure, the ICS will not fulfil its promise in all response efforts, and that a diverse and complex array of conditions have to be in place pre-disaster and during a response for the system to work as designed.

This idea implies that the standardized systems like the ICS, among others, used by firefighting and oil spill response agencies, and the maritime SAR

IAMSAR plan systems used within the maritime rescue co-ordination centres do not come in a 'one-size-fits-all' package. Adjustments might have to be made.

Bechky and Okhuysen (2011) discuss how organizations facing uncertainty can engage in organizational 'bricolage', restructuring their activities by role-shifting, reorganizing routines, and reassembling their work. Moynihan (2009) emphasizes that switching between centralized-hierarchical and more decentralized, network-based governance forms can be appropriate. The presence of norm-based, professional, or situation-based trust in the skills, dedication, and commitment of the participants can allow for higher system complexity without centralized authority (Borch, 1994; Beck and Plowman, 2014). With a high degree of mutual trust in the other agencies acting to achieve a common goal, the strict hierarchical control and command structures can be loosened.

Wolbers et al. (2018) advocate multiple co-ordination sequences with ad hoc adaptations. Each sequence should allow for postponement of co-ordination where disintegration and fragmentation will facilitate sharpened team identity and a faster-acting, as well as flexible and creative role execution. Beck and Plowman (2014) suggest that 'deploying portions of a portable structure' such as the ICS can bring in enough structure to create some order, but not inhibit experimenting, a search for improved and adequate solutions, and a self-organizing process.

Flexibility and Structuring Mechanisms

Organizational structure may be seen as a formal configuration of roles and procedures. Several structuring tools are available to shape interaction and create a certain amount of control. These include centralization-decentralization, the vertical and horizontal span of control, functional specialization, formalization, and autonomy. Ranson et al. (1980) regard organizational structure as 'a complex medium of control' which is continually shaping interaction. At the same time, the organizational structure is influenced by this interaction. A high degree of centralized control can hamper the flexibility of the organization. This control can also be difficult if several otherwise independent agents are involved in a joint action. To achieve a balance between hierarchical control and autonomy, Bigley and Roberts (2001) introduce four basic processes for adapting an emergency response structure to a complex task environment. These are structure elaboration, role-switching, authority-migrating, and system-resetting.

A central element in a joint emergency response system is to allow for an enlargement of the joint organization through adding new elements. Bigley and Roberts (2001) describe this as *structure elaboration* where the system is constructed with more units assembled, positioning them within the present structure assigning roles, tasks, and procedures to the newcomers. This approach can also call for *role-switching*: assigning personnel to different positions within the organization according to needs and qualifications. An important element to create flexibility is *authority migration*, where persons with special qualifications are given the freedom to perform tasks that are not dedicated to their role. This relocation of tasks and resource control will also influence other persons' role span and can create internal tensions. Even more challenging is the fourth structuring mechanism suggested, i.e. *system-resetting*. The changes in the composition of the system can refer to withdrawing an organization from specific tasks, moving units to other tasks, discharging units, and changing the couplings between the units.

The Figure below illustrates the dimensions related to the structuring of emergency response systems to create the necessary co-ordination.

Figure 12.1 shows how the context of the operation in the form of task complexity can call for structuring mechanisms that reflect increased task complexity. We claim that the lead organization managing the joint operation and linking the organizations together has to incorporate structuring mechanisms that can adapt the organizational form and provide new patterns of managerial roles. The structuring mechanisms have to facilitate the allocation and configuration of managerial roles in an optimal way to achieve the necessary co-ordination over time.

The complexity of value chain configuration in large-scale joint operations calls for the adapted managerial roles and mechanisms within the co-ordinating organizations. The lead organization of major disasters and high-complexity environments are in need of these mechanisms for ad hoc adaption of the incident command roles.

Methodology

Research Strategy

This study builds upon in-depth inductive case studies of two challenging search and rescue operations in high-complexity environment. The cases

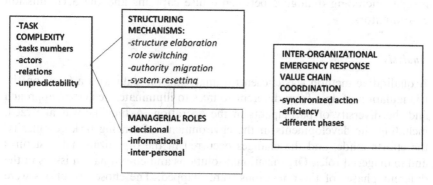

Figure 12.1 Analytical model

were chosen because of the presence of several unpredictable factors related to the tasks, owing to challenging conditions such as bad weather, extreme cold, and darkness. The cases also illuminated the extra challenges of maritime operations. The cases provided rich material, with several aspects of relevance to managerial roles and structuring challenges of joint rescue operations.

Data Collection

The data was collected through several methods:

Document studies: A broad range of documents was studied. This included logs and reports from the involved agencies and newspaper and magazine articles. Official reports from the government examination boards were important for fact-finding and evaluation.

Interviews and panels: The data collection included the following panel discussions and presentations: Exercise Nord at Nord University, Bodø, Norway, 23 March 2019; Fourth Joint Arctic SAR TTX of Association of Arctic Expedition Cruise Operations (AECO), Reykjavik, Iceland, 9−10 April 2019; Emergency Prevention, Preparedness and Response (EPPR) Working Group Meeting Bodø, Norway 3−6 June 2019; Civil Protection Conference, Oslo, Norway, 3−4 February 2020. Also, in-depth interviews and discussions with SAR mission co-ordinators and Joint Rescue Co-ordination Centres (JRCC) directors were performed on specific subjects. These interviews were made as follow-ups of the documentary studies to provide more insight into the organizational adjustments made.

Tabletop and simulator exercises: Different scenarios related to possible trajectories, primarily related to worst-case scenarios, were tested in NOR-DLAB Nord University Emergency preparedness laboratory. Part of these tests was made as an exercise with professional emergency response co-ordinators taking part in a course for on-scene co-ordinators. Also, participant observation arranged by two JRCCs and the AECO gave valuable insight, including dialogues between cruise captains and the SAR mission co-ordinators.

Analysis

A qualitative method was chosen for analyzing the data with a focus on the triangulation of data from different sources to illuminate the actors' approach and the diversity of participants in the operation. A timeline was analyzed, including the developments in the environment influencing task complexity, the efforts made, and the changes occurred in the organizational structures and managerial roles. Organizational solutions and co-ordination issues in the different phases of the operations were mapped. The chosen solutions were underlined by citations from the key personnel involved.

Data

Institutional Context

The structure of the SAR system in most countries is divided into sub-functions and action sectors, for example, according to incident types, geographical area of responsibility, and capacities within the different levels of co-ordination. For maritime SAR operations, the international IAMSAR Manual serves as a platform as to tasks and roles. The Norwegian SAR system is organized with a National and a Regional structure. The two Joint Rescue Co-ordination Centres (JRCC) are responsible for SAR operations at the national level. Operations are co-ordinated either from the two main centres in Stavanger (Sola) and Bodø or are delegated from the JRCC to one of the 13 rescue sub-centres (RSC) managed by the regional police: one in each police district at the regional level. The chief of police runs the regional strategic level in each police district. In major incidents, the police incident commander onshore will be supported by incident commanders from the paramedics, the fire brigade, civil defence, and volunteer organizations. The incident commanders will form a command centre on the scene, with the on-scene police incident commander in charge of the shore operation. He will report to the RSC.

International conventions and standards for maritime and aeronautical SAR services are set by the International Maritime Organization (IMO) and International Civil Aviation Organization (ICAO). The International Convention on Maritime Search and Rescue (IMO, Hamburg Convention) and the Convention on International Civil Aviation with its Annex 12 (ICAO, Chicago Convention) provide the rules and regulations for SAR services. The International Aeronautical and Maritime Search and Rescue Manual (IAMSAR Manual), published by the IMO and the ICAO contains procedures for the organization of maritime and aeronautical SAR, mission co-ordination, operations of search and rescue units (SRUs), and provision of SAR-related training. The Manual is not binding but provides an internationally accepted foundation for the appropriate provision of maritime and aeronautical SAR services. The search and rescue mission co-ordinator (SMC) will be in charge of the overall co-ordination of incidents and allocation of all necessary resource. SMCs are rescue controllers who work at the JRCC.

In large-scale operations, the SMC usually appoints an on-scene co-ordinator (OSC), who operates on a tactical level co-ordinating SRUs, aircraft, and other assisting units or vessels arriving at the incident site. The OSC is responsible to the SMC and reports to the JRCC during an incident. The OSC is usually the captain of the first SAR unit arriving on scene until the SMC relieves the OCS from his/her duties. This unit can be either a professional SAR unit, for example, a vessel, aircraft, or land-based team (land SAR) from the coast guard, coastal administration, armed forces, fire and rescue services, police, The Norwegian Sea Rescue Society, the Red Cross, or a private unit such as a vessel or aircraft that is the closest to the incident with a capability to handle OSC duties. The OSC should be the most capable person available to fulfil the responsibilities. According to the SAR co-operation plan between SAR services and passenger ships in an emergency, the OSC has the following responsibilities:

- Assume operational coordination of all SAR facilities on scene
- Receive the search action plan from the JRCC
- Modify the search action plan based on prevailing environmental conditions and keeping JRCC advised of any changes to the plan
- Provide relevant info to the other SAR facilities
- Monitor the performance of other units participating in the search
- Coordinate safety SAR facilities involved.
- Make consolidated reports (SITREP) to the JRCC, and report destination, number and names of survivors aboard each unit and request additional assistance from JRCC when necessary, such as medical treatment/ evacuation of seriously injured survivors (JRCCs Northern and Southern Norway, 2019).

An aircraft co-ordinator (ACO) can be appointed by the SMC to co-ordinate aerial units arriving at the incident site. The ACO activities aim at the effective co-ordination of aircraft and contributing to aeronautical rescue and SAR services. The ACOs follow highly developed routines for executing their tasks (JRCCs Northern and Southern Norway, 2019). ACO in Norway would most likely be situated either at a Coast Guard vessel or on board one of the Air Force planes such as the surveillance patrol planes or helicopters.

At an incident site, the captain of the distressed vessel plays a major role in on-scene management. The captain of the distressed vessel is responsible for the vessel's and passengers' safety for all types of acute emergency and preparedness incidents in which the vessel is involved.

The shipowner also has its own crisis response team to handle different tasks that the company is responsible for during incidents. The company crisis response organization should be able to assist the SAR services by organizing support, equipmentand liaison with any of their vessels and provide relevant information for the SAR services.

Even though a large number of new vessels are designed according to the high standards and built tailor-made for the navigation in the polar waters, accidents can still happen. An itinerary through remote, non-populated regions could call for extra safety and security precautions, including emergency preparedness. Accidents like fire on board a vessel, collisions, and grounding of larger vessels in polar waters are among the most difficult tasks for the emergency response systems (Borch, et al. 2016a, b). The main challenge for preparedness system capacities for tourist vessels is managing the response to a large number of people, having enough rescue equipment, speed of mobilization, and good co-ordination (Borch et al., 2016c; Andreassen et al., 2018b).

Case 1: The MV Viking Sky SAR Operation

On 23 March 2019 *MV Viking Sky*, a conventional cruise ship, was south-ward bound to Stavanger in the inner shipping lane of the Norwegian west coast. This was the last of six Northern Light cruises to northern Norway and their first Northern light season. The weather was fierce as the *MV Viking Sky* crossed the shallow and treacherous sea area Hustadvika between the coastal towns of Molde and Kristiansund.

At around 13:50 a blackout in the engine occurred on board *MV Viking Sky*. The main generators, which deliver electricity to the main engines, stop-ped. The vessel lost propulsion power and started drifting towards the shore. There were 1,373 passengers and crew on board. The 915 passengers were mainly from UK and the USA. The crew of 458 came from 45 nations. A strong gale wind was blowing with reported gusts up to 43 mph and waves of some 8–10 metres.

The captain instructed the crew to drop both anchors. However, the anchors did not hold, and the ship continued to drift astern towards the shore at a speed of 6–7 knots. At 14:00 the captain of *MV Viking Sky* sent a 'mayday' signal from position 63.003°N, 6.596°E (JRCC Norway, 2019). It was then off the coast of Hustadvika near Fræna, a municipality in More og Romsdal, a county of around 261,500 people in western Norway.

In this area the JRCC South Norway at Sola, Stavanger is responsible for the SAR operation. The mayday call was received by Maritime Radio South and was responded to by the JRCC SAR mission co-ordinator on duty.

The SAR mission co-ordinator within the first 15 minutes alerted the local vessels including the rescue cutter through the radio operator at Maritime Radio South, then the SAR helicopter resources and the Rescue Sub-Center (RCS) at the More and Romsdal Regional Police (112).

At 14:09 The Norwegian Coastal Administration was informed and started preparing for an oil spill response operation.

At 14:13 the captain of *MV Viking Sky* activated the general alarm on board, and the passengers and crew began to go to muster station. The master considered evacuating passengers and crew into the lifeboats. How-ever, given the environmental conditions, this was considered to be too dangerous.

Some helicopter resources were a mere hour away. Other helicopters were mobilized from other regions of Norway. SAR helicopters from nearby oil and gas installations were scrambled. In total, six helicopters were involved in the operation.

The JRCC SN mobilized more SAR mission co-ordinators. The JRCC National Rescue Board was called in with representatives from all the important agencies with resources, including liaisons to others government agencies, such as the Norwegian Coastal Administration, the military, and the aviation authorities.

A dedicated person from the Avinor air traffic control centre arrived at the JRCC and took the role as the ACO communicating at the VHF air channel with *MV Viking Sky*, the helicopters, and the OSC at the coastguard vessel.

The shipowner company set the preparedness team, creating a next of kin telephone service. The insurance company supported both the shipowner and the JRCC in deploying towing boats that could help the vessel out of the shallow waters. The towing vessel with sufficient capacity was 12–15 hours away situated in Bergen.

Furthermore, the Norwegian Coastal Administration contributed to the operation through their Vessel Traffic System (VTS). The two pilots from the Norwegian Coastal Administration (NCA) pilot service on board *MV Viking Sky* provided important information. The two pilots served as a liaison between the *MV Viking Sky* captain, the JRCC, and the coastal administration.

At 14:39 a local incident command post (ILKO) was established. The Norwegian Red Cross got an alarm call from the police and quickly put out an alert to 400 of their volunteers along the west coast and central Norway. They organized a reception centre in Bryne sports hall together with the local fire brigades and local volunteers. Sixteen ambulances and 40 buses were mobilized, and four hotels were prepared for accommodation. Evacuation centres and next of kin centres were established in the town of Molde and Kristiansund. The fuelling of the helicopters was taken care of by the local fire brigade and police close to the sports hall using tank trucks from Molde airport. In total, almost 500 people were involved in different parts of the response value chain.

At 14:29 (reported to JRCC at 14:44) the *MV Viking Sky* had one engine up and running. Together with the anchors, they managed to keep the bow up to the wind and stop the drift. The ship was only about 100 metres away from the shallows and had, during the drift, passed close to or over a 10 metre shoal. The draft of *MV Viking Sky* was 6.65 metres.

At 15:00 the first vessel arrived on scene. However, owing to the wind and waves, they could not support the *MV Viking Sky*.

At 15:03 the first helicopter arrived and started airlifting. The first wounded passengers were delivered to the reception centre at 15:40. In total, 30 helicopter loads were performed with six helicopters over 18 hours, evacuating 460 passengers. The number of rescued in each load varied from two to 23 persons.

At 16:13 the Norwegian coastguard vessel *Njord* took the OSC role and was on the scene at about 16:40. The vessel was unable to help with evacuation because of the weather, but took care of the co-ordination of other vessels on the scene.

The first tugboat arrived at about 16:40. However, the weather conditions were too severe to secure a towline.

At 20:30 the status was that five vessels were on standby in the vicinity and three towing vessels and three coast guard vessels were on their way.

At 20:55 the JRCC called the Rescue Co-ordination Centres (RCC) in Sweden, Denmark, and the United Kingdom and requested assistance according to the Host Nation Support guidelines. A Danish SAR helicopter was sent to Kristiansand to serve as a back-up in the southern region, and a Swedish helicopter was on standby to support the Oslo region.

Four hundred and seventy-four people were evacuated by helicopters. The extreme wind conditions made the operation complicated. There was a risk for the vessel to ground and capsize, with the potential for a major disaster where people could fall over board in the cold water. The majority of the passengers were 60 years of age or older, increasing the risk of severe casualties. The operation was, therefore, upscaled as fast as possible. Another 800 persons were preparing for a new phase if the *MV Viking Sky* grounded. The Red Cross had 40 shore rescue specialists on standby. The crisis staffs were mobilized in the three neighbouring municipalities to prepare for medical treatment and evacuation.

At 19:42 another rescue operation had to be arranged, as a cargo vessel crossing Hustadvika also had a blackout and sent out a mayday call. Two of the SAR helicopters were allocated from the *MV Viking Sky* operation to rescue the crew of nine persons from *MV Hagland Captain*.

On the next day at 08:00, towing was established with two tugboats, and after one hour they managed to slowly turn the vessel and bring it out in the open sea and towards the town of Molde.

At 09:11 the JRCC terminated the helicopter rescue operation. At 13:09 the RSC command centre at the sports hall was closed and the police resources allocated. At 15:11 the mayday for *MV Viking Sky* was cancelled by the JRCC after consulting with the captain after an SAR operation of more than 24 hours, and the resources were de-mobilized. *MV Viking Sky* was guided into Molde harbour and moored at 16:25.

The *MV Viking Sky* SAR operation was the biggest evacuation operation by air from a vessel in Norway. Three persons were seriously injured, and approximately a dozen were hospitalized. There were no fatalities. Six helicopters, 17 ambulances, 40 buses, three municipalities, five hotels, and over 300 volunteers in addition to professional emergency response agency personnel were involved in bringing the passengers to shore and in to the evacuation centres.

Intra-case Analysis Case 1: MV Viking Sky

One of the challenges was the understanding of the seriousness of the situation and the upscaling of the operation. It was difficult to create a picture of the worst-case scenario and how fast the situation would escalate towards a grounding and evacuation situation with more than 1,000 persons in lifeboats, life rafts, and in the sea. It took some time before the police mobilized for a major disaster at the operational and strategic level.

> We did reflect on the worst-case scenario, but it was difficult to see what could be done if the vessel capsized.
>
> SAR mission co-ordinator

Communication in the different part of the value chain was challenging, establishing a collective situational awareness and co-ordinating activities between the actors. Maritime coastal radio, portable VHF radios for communication with helicopters, and the Tetra emergency network emergency network were in use. Some units were not equipped with all communication tools. This caused uncertainties about the wounded persons coming in.

> We did not know the conditions of the passengers that was brought in before they arrived. This was frustrating and may have caused delays before the necessary treatment.
>
> Representative from the regional Red Cross organization

At the JRCC they acknowledged the need for more advisers to call upon in response to larger incidents. Furthermore, the resource registers for different maritime resources such as towing vessels could have been better. It took some time before the regional police acknowledged the incident a major catastrophe situation establishing more operational capacity at the staff level and evacuation centres and informing the County Governor responsible for the overall long-term resource allocation.

> An important lesson learned was that the police should have established their emergency staff and mobilized the RSC regional rescue board at an earlier stage. If the potential disaster had manifested, the police and other response agencies may have been delayed in their response.
>
> Evaluation of the *MV Viking Sky* incident. Report from DSB (Norwegian Directorate for Civil Protection)

The logistics in all parts were a great challenge because of the number of units that had to be mobilized from different parts of the country. Among other things, because of the accident happening at the weekend, mobilizing and transportation of more helicopter crew to the site was a difficult task. For the JRCC, informing all stakeholders and receiving the necessary information was a challenging task. They had to upscale the teams and prepare for a long-running operation. It was also difficult to reflect on how the incident would develop and what scenario to prepare for with configuration of the value chain. The activity at the staff level here was critical. Added capacity and more actors involved, as well as ways of structuring these joint activities, were important issues, as the operation could take a long time and include many casualties. Among other things, an expanded, co-ordinative role of the National Rescue Board was discussed.

On a large scale, mass rescue operations calling for huge resources and co-ordination at national level became an important issue both for information exchange and resource allocation. Information on the political levels and co-ordination at ministry and directorate level were challenging. This also included the regional county governor administration. The Norwegian Directorate for Civil Protection (DSB) has a role here. However, it has not been practised. The DSB's efforts to improvise and arrange co-ordination meetings were regarded as disturbance from the JRCC.

> Norway has not developed a national plan for mass rescue operations. The JRCC recommends the development of such a plan. It will create a platform for interaction between the JRCC and other agencies...
>
> Evaluation of the *MV Viking Sky* incident. Report from DSB
> (Norwegian Directorate for Civil Protection)

The master was responsible for what was happening on board. He announced that he had a plan for the rescue operation. However, he did not communicate and discuss this plan with the JRCC.

> An evacuation was an extremely difficult decision to make, and more people with specialized knowledge of the terrain and survival in the heavy seas should have been involved.
>
> SAR mission co-ordinator

The operation could have been going on for a very long time. A difficult decision for the managers of every agency involved and especially the JRCC and RSC/chief of police would be to balance and add the resources into the operation, giving the first responders time to rest. As an example, all but three persons at JRCC South Norway were engaged in the first hours of the operation. In total, 15 SAR mission co-ordinators were active during the operation.

Case 2: MV Northguider

On 28 December 2018 the Norwegian fish trawler *MV Northguider* was fishing for shrimps in the Svalbard archipelago when it grounded. The *MV North-guider* had set out from mainland Norway almost two weeks earlier and was now trawling the northern part of the Hinlopen strait between the main island of Svalbard, Spitsbergen, and Nordaustlandet (North East Land), at 79.53°N, 18.4°E. There were no other ships in the vicinity (JRCC North Norway, 2019).

A north-westerly gale-strength wind was blowing with heavy snow showers. At this time of the year, there is limited daylight, so the grounding happened in complete darkness. The wind and the freezing temperature of minus 22°C forced the crew to remove ice from the deck to maintain workplace safety and vessel stability. After turning the vessel to a northern course in the north of the Hinlopen strait, the wind and current took the ship and drove it towards the shoreline. The force of the heavy trawl made manoeuvering difficult.

At 13:00 the captain lost control of the vessel, and the *MV Northguider* grounded. The engine stopped, causing loss of electricity and a blackout on board. The crew of 14 gathered on the bridge. The vessel rolled heavily when the waves hit the hull. It was therefore difficult to keep balance in the wheel-house, and the crew struggled to put on lifejackets.

At 13:22 20 minutes after the grounding, the captain and first mate sent out a distress signal on the maritime radio HF/MF DSC-Digital Selective Calling. They also released the satellite-based EPIRB-Emergency Position Indicating Radio Beacon transmitting the vessel Maritime Mobile Service Identity (MMSI) number and position.

At 13:22 the DSC distress signal and the EPIRB-satellite signal were received by the JRCC North Norway (JRCC-NN) and the Coastal Radio North in Bodø. A mayday relay was sent to all ships. The radio operator at Coastal Radio North tried to reach the *MV Northguider* via MF-maritime radio but did not succeed communicating with *MV Northguider* (HF-maritime radio with a longer range was closed down at Svalbard in 1999). In the meantime, the captain of *MV Northguider* had contacted the shipowner via Iridium satellite telephone informing him about the critical situation and the need for immediate response. He then called JRCC-NN informing it about the distress and the number of persons on board. The SMC at JRCC-NN tried to reach *MV Northguider* via Iridium satellite telephone without success.

The crew released one of the life rafts, but the line broke in the strong wind, and the life raft disappeared. They now had only one life raft left, and this was difficult to release because of the heavy listing. Thus, the situation became more time-critical, without the crew being able to communicate with the rescue co-ordinators.

At 13:28 the operational centre at the County Governor's office in Longyearbyen was informed and asked to mobilize the two SAR helicopters at the airport.

At 13:32 both SAR helicopters at Longyearbyen were scrambled.

At 13:33 the police incident commander travelled to the helicopter base.

The JRCC checked out the position of other vessels in the region through the Automatic Identification System (AIS). However, the closest ship was 18 hours away from the distressed vessel. The JRCC was in contact with the closest coast guard vessel *Barentshav* located near Bear Island in the Barents Sea approximately 24 hours away.

From the ice charts, an additional problem was emerging as the ice was pushed down to the Hinlopen strait. This situation called for icebreaker assistance. However, the only coast guard vessel with ice breaker class was at the mainland base and the crew away for the Christmas holiday.

The SAR mission co-ordinator also asked the Norwegian Joint Head-quarters of the armed forces for support from the Orion maritime patrol air-craft at Andoya to serve as a communication link and helicopter support (air co-ordinator). They also discussed landing at Longyearbyen and loading rescue drop kits if the crew was stranded for a longer time. However, the Orion had to go to base for fuelling. They therefore did not take off before 15:20 from the mainland. At that time the helicopters were already on-scene.

The media was informed about the incident on Twitter and the JRCC web page. The need to give information to the next of kin and the public was pressing. The JRCC issued a press release and took care of the increasing media interest through their media adviser.

In Longyearbyen, the police force at the County Governor's office mobilized the hospital and arranged for the crew to be taken care of upon arrival in Longyearbyen. Longyearbyen Red Cross arranged a reception centre in Longyearbyen, and the shipowner established a next of kin call centre at its headquarters on the west coast of mainland Norway.

At 13:40 the mate on board *MV Northguider* managed to establish a MF radio communication with the *NoCGV Barentshav* serving as a relay station between the JRCC and *MV Northguider*. *NoCGV Barentshav* could inform the crew on board *MV Northguider* that the helicopters would arrive in 40 minutes (sound log).

The captain of the distressed vessel informed the coast guard that the conditions on board were getting worse. The crew had to move out of the wheelhouse to the deck to avoid being trapped if the vessel capsized. The last life raft was stuck under the vessel because of the increased listing. If the crew had to leave the vessel now, they would have to jump and swim in the icy water towards the ice flow at the beach around 50 metres away. This situation would be very risky in the heavy sea with ice and Polar bears nearby. The crew were waiting at the deck for approximately half an hour. The emergency generators that started after the main engine stopped also failed, and the crew was forced to sit in complete darkness. They tried to keep their spirits up by singing.

At 13:50 Longyearbyen hospital was alerted. The hospital called for an ambulance plane from the university hospital on the mainland. Longyearbyen Red Cross was also mobilized about the need for personnel and resources. It prepared SAR drop kits to be launched by the Orion patrol plane or other aircraft.

At 13:58 the Governor of Svalbard as head of the RSC was informed and mobilized the police staff.

Around 13:57, some 35 minutes after the alarm call from the JRCC, the first Super Puma SAR-helicopter took off from Longyearbyen with a crew of six, including a doctor and the SAR technician (SAR Tech). The second helicopter followed 15 minutes later. The helicopters were fuelled for a 3½- or 4-hour operation, as the pilots were informed that the crew might have entered the life rafts, meaning that a long search might be needed. The SAR helicopter crew did not have much information about the conditions of the persons in distress. If the crew of the distressed vessel had to jump in the sea, they would suffer from hypothermia in a very short time and might drift away in the strong wind and waves, requiring a time-consuming search operation. The risk of injuries would also be significant.

Owing to the communication challenges, the RSC at the Office of the Governor of Svalbard agreed with the JRCC SAR mission co-ordinator that the RSC should handle the communication with the helicopters.

If the crew was seriously injured and required treatment before they were brought to hospital, they would have to be brought to an abandoned research station in the vicinity and treated under the primitive and cold conditions there. The crew might have to be brought onto the beach first.

The accident happened 105 nautical miles from Longyearbyen. The Super Puma has a maximum range of 260 nautical miles under ideal conditions. With heavy wind and fully loaded, the range is reduced. Also, the time hovering when lifting the crew would reduce the range. If only one helicopter had been available and the helicopters had to spend time searching for persons in the water, they would have to refuel from one of the fuel depots on the archipelago. There was an extra depot located at Kinnvika not far from the scene. However, this fuelling process could be hampered by bad weather, including ice and snow, and take extra time.

At 15:06 the JRCC reported to the RSC that the military patrol plane Orion had left the mainland and expected to reach Longyearbyen after approximately two hours and load SAR drop kits. The JRCC also informed the RSC that if the crew had to jump in the sea, they would have to swim 50 metres to reach the ice ridge. They did not know what it would be like to enter the ice and to reach the beach from the ice flow.

At 15:15 the first helicopter was on the scene after 45 minutes in the air. The snow and total darkness made it difficult to find the vessel, however. The pilot and the vessel captain communicated through the portable VHF radio, and the vessel sent a flare showing the location of the vessel.

Owing to the weather, the pilots had problems finding the reference points when approaching for lifting and had to use some time on manoeuvering.

Owing to the weather conditions on the scene and the critical situation for the crew, the Governor of Svalbard in his evaluation report summarized the operation as follows:

> Under the circumstances with the weather conditions on-scene, the helicopter crew balanced close to the limits to save lives.
>
> Evaluation report, Governor of Svalbard, 2019

The paramedics and SAR technician were launched down to the deck and informed the crew about the procedures. They then started lifting the crew members. Ten of the crew were rescued in the first helicopter. The last four persons had to wait for the second helicopter to arrive. By this time, they had lost communication with the helicopters as the portable VHF radio was out of power. The batteries did not last long in the cold weather.

At 16:19 the second SAR helicopter reported that the final crew had been rescued. The operation of lifting the crew of 14 with the two helicopters took approximately one hour.

The whole operation took approximately four hours from the time that the distress signal was sent from *MV Northguider*.

At 17:00, after approximately three hours in the air, the helicopters were back in Longyearbyen. The Longyearbyen fire brigade was mobilized to provide transport to the hospital. The crew was taken care of by medical personnel at Longyearbyen hospital.

A large mobilization had taken place with civilian, military, and voluntary resources. At national level the JRCC made situational reports to the the the government crisis support unit at the Ministry of Justice and Preparedness in Oslo.

Intra-case Analysis Case 2: MV Northguider

The SAR operation took place under challenging weather conditions. The situation for the crew was fast deteriorating. There was uncertainty about the crew members having to jump in the water and whether they would make it to shore and manage to climb the ice on the beach. If they floated away, a search operation would have been necessary amid terrible visibility, darkness and heavy waves. The survival team for the crew in the water would have been limited.

> I thought my last hour had come. The weather was bad, and we did not
> know if the vessel would capsize in the next second.
>
> Chief mate, MV Northguider

For the helicopters, an extended search operation would have meant refuelling at the nearby depot. That would have delayed the operation.

> The weather is bad on the scene. Heavy snow and very low visibility.
> Wind force up to 30 knots and minus 25 degrees Celsius. It will be very
> difficult to get into position, but we are preparing for hoisting. It is not
> possible to inform the SAR mission co-ordinator through satellite
> because of time limitations and the weather conditions demanding full
> pilot concentration.
>
> Pilot SAR helicopter 1

The radio communication challenges made information exchange a problematic issue. The SMC at the JRCC had to delegate authority to the Rescue Sub-Center Svalbard and police at the County Governor office in Longyearbyen serving as a mission co-ordinator directing the helicopters. Also, the pilots were on their own, having to co-ordinate the whole operation by themselves. As a result, the four pilots served as both SRUs, OSC, and ACOs at the same time.

The Rescue Sub-Center (RSC) took care of the reception and the mobilization of resources locally, helping out the JRCC with the next step in the operation. At the start, it was not made clear that the JRCC should co-ordinate the whole operation, nor was agreement reached on role-sharing and tasks as usual in SAR operations at sea. However, it was agreed that the contact with hospitals and the SAR units should go through the JRCC. Owing to capacity limitations on the scene, the task of communicating and discussing the situation with the helicopter pilots was delegated to the RSC incident commander. He thus took over some of the co-ordination and communication tasks of the SAR mission co-ordinator. The operational leader was familiar with the area, the communication challenges, and what the available alternatives were. He could, therefore, provide the pilots with the necessary advice and give the pilots time to concentrate.

It was agreed that the JRCC would take care of media and other information. The shipowner took care of next of kin after discussing with the RSC.

Analyses and Discussion

Co-ordination in High-complexity Environments

The cases above illustrate the managerial challenges of co-ordinating emergency response operations in a complex task environment. The number of difficult-to-predict factors was high in both cases, making situational awareness a difficult task, especially about the worst-case scenario. Several trajectories of development and finding effective response create a headache for the managers in charge.

The context overview above shows that the weather was bad, causing many challenges for the SAR operation in both cases. In the *MV Viking Sky* case, the proximity to the shore and the risk of grounding and capsizing caused the greatest concern. In the *MV Northguider* case, the risk of capsizing was imminent, with limited resources available for search, rescue, and medical support.

In the case of *MV Viking Sky*, the uncertainty about what would happen after a grounding and the number of potential casualties called for an unprecedented scale of operation. A vast number of personnel and units from government, private companies, and volunteer organizations had to be mobilized and co-ordinated, causing a ripple effect for other regions. The dependency on specialized capacities such as emergency towing vessels and a large number of helicopters called for extra co-ordination efforts, including sufficient air control expertise.

The potential worst-case scenario of the cruise vessel grounding and an adjoining evacuation in extremely heavy waves of a thousand persons in the sea,

Table 12.1 Inter-case analysis of the operational context

	MV VIKING SKY	MV NORTHGUIDER
Wind	Southwest strong gale to storm winds Beaufort 9–10m/second or 22–25 m/second	Strong gale up to 30 knots
Waves	Total significant wave height over deep water 9–10 metres from west	High
Visibility	3km. Occasional snow showers	Very low. Snow showers. Arctic darkness all day
Temperature		Very low – 26°
Distance to shore	A hundred metres from grounding	Grounded close to the beach
Regional SAR capacity	Very high. The offshore fleet at sea. Private and government helicopters. Significant rescue agency resources	Limited to local resources at Longyearbyen. Two helicopters
Radio communication	All communication channels available	Limited. MF only through relay station on coast guard vessel
Physical communication	Very good on land, sea, and air	Very limited
SAR Helicopter units	6+	2
SAR vessel units	Numerous	none
Casualties	20 wounded	No casualties
Persons in risk	1,373	14
Distance from base	Low	Long
Distance to hospitals	Low	Long

in rafts and lifeboats, increased the uncertainty about the tasks and trajectories. A vital question was about what the adequate strategies and resources were, and when the resources should be mobilized. As one stakeholder said,

'It was difficult to see what the worst-case scenario was to predict and plan for it'.

SAR Mission Co-ordinator JRCC-SN

The unpredictability of emergency response and the need for flexibility was highlighted when another emergency came up. The mayday call from a cargo vessel where nine people had to be evacuated came on top of it all and called for additional efforts from the units, especially the helicopter crews. One important aspect was the staying power of the crews and helicopter in an operation that could go on for a very long time.

The weather was a large, unpredictable threat in the *MV Northguider* case too. The wind and snow made the helicopter search and lift operation a challenging task for the helicopter crew. If the vessel capsized and the crew of *MV Northguider* had to jump in the water, they would be difficult to find, and all would soon suffer from hypothermia and require immediate medical help. Alternatives to moving the rescued persons to the abandoned research station for first aid were considered.

The challenges that the helicopter crew were summarized as follows from one of the pilots of a SAR helicopter: '*This task was demanding, but this is for what we are trained for*'.

Communication was a very challenging issue. First, it was difficult to inform the crew about the rescue operation, and second, it was hard to get a situation report from the scene. Thus, there was considerable uncertainty about the rescue context. The darkness and snowstorm caused difficulty with locating the distressed vessel upon arrival from the helicopters. Even though the helicopters had back-up depots nearby, the operational limits, owing to a finite supply of fuel, was a present factor. Fuelling would take precious time from the SAR operation.

For both cases, the large number of factors that could go wrong made the planning process a challenging task. Co-ordination was complicated in the *MV Viking Sky* case by the number of units involved, the problems with the availability of adequate towing capacity, and rescue operations close to the shore and an additional SAR operation caused by the cargo ship in distress. In the *MV Northguider* case, the limitations of available resources other than the two helicopters, the communication challenges, and the severe weather conditions made co-ordination a challenging task, with much of the responsibility placed in the hands of the helicopter pilots.

The number of stakeholders in a major accident will usually be overwhelming. The *MV Viking Sky* case showed that multiple nationalities of the passengers and crew might be an issue in the Arctic areas, where the crew alone came from 30 different countries. The vessel owner might not be the managing operator. In addition, a third flag state might be a third stakeholder, and an insurance company from another country still. Keeping all parties updated requires a lot of resources:

> We considered using the National Rescue Board for communication purposes to release the mission co-ordinators.
> SAR mission co-ordinator, MV Viking Sky incident, JRCC-SN

The discussion above has shown that in the Arctic, emergency response conditions can quickly become highly complex, and the capacity will be stretched to its limits. The weather conditions could severely worsen the situation for the units in distress, and darkness, snow, and ice can hamper the rescue operation. Finally, there will be limited availability of adequate resources, where the incident commanders have to rely on other types of resources not tailor-made for the tasks at hand.

Task Complexity and Managerial Roles

Large-scale SAR incidents in the Arctic can result in the need for strengthening capabilities of the system to respond, as the roles of the co-ordinators could vary because of the lack of resources and lack of experience in these kinds of incidents in the Arctic. The involvement of more management levels could be challenging and time-consuming, as well as causing informational challenges. Flexibility in the decision-making processes is important at all management levels, including finding new resources and solutions, as well as adapting standard operating procedures to the prevailing environment.

The on-scene co-ordinator role is very challenging and often requires a whole team around the person who is assigned as an OSC. In large-scale maritime operations, the OSC is an essential communication link between the JRCC, the distressed vessel, and the units arriving on the scene. Communication between RCCs, industry, and crew could be challenging for co-ordination. On-scene co-ordination could be complicated by language problems, panic, reduced communication coverage, and wind, so establishing the structure, like a good operational communication in possible complex conditions, and routines on how to proceed, could be necessary. Therefore, the decisional roles could be most the important for the OSC, especially in the event of uncertainties and limited situational awareness.

Table 12.2 shows how the different tasks were performed at different management levels and the mobilization of the standard maritime SAR personnel, including added capacities at the strategic level following the Norwegian SAR system regulations. The managerial roles within sea rescue operations follow strict regulations based on the IAMSAR Manual for sea and air rescue. Each country has one or more maritime rescue co-ordination centres with specialized SAR mission co-ordinators. They delegate authority to captains at Samaritan vessels on the scene, preferably coast guard vessels as they are specially trained in the role as on-scene co-ordinators or incident commanders, with special tasks of searching for survivors and rescuing people. In the *MV Viking Sky* incident, the captain on the arriving coast guard vessel was appointed as an OSC. However, even though five or six vessels were present, they were not able to take part in the rescue operation, owing to the waves and proximity to shore.

Instead, the police were in charge of the land-based operation from their shore command centre. The ACO role is typically given to the coast guard in maritime helicopter rescue operations. However, no coast guard vessel with ACO competence was in the vicinity. There was a large number of helicopters that had to be co-ordinated. The ACO management role was given to a regional air traffic controller from Stavanger airport that was located at the JRCC with support from the regional Møre aviation terminal control centre.

Table 12.2 Managerial roles within the standardized SAR system

	MV VIKING SKY	MV NORTHGUIDER
On-scene co-ordination	Broad involvement from all emergency agencies onshore and offshore (coast guard)	Limited. The four pilots had to co-ordinate all activity on-scene
Air co-ordination	Taken care of by air traffic ontrol	Co-ordinated by pilots
SAR mission co-ordination	Several SAR mission co-ordinators. Added administrative resources	Limited involvement owing to lack of communication Local incident commanders at the County governor office
SAR mission communication	Maritime Radio South and mobile phone	MF radio with coast guard Limited satellite communications
Oil spill response capacities	Very high	Very limited
Media communication	Well developed through JRCC media contacts and all media	Very limited. County governor and local newspaper
Strategic management	National SAR board mobilized National oil spill response authorities mobilized	National SAR board was not mobilized

At the regional operational level, the police did not mobilize their staff at the beginning of the operation but added more personnel at the operational centre. This circumstance gave the operational leader a much broader responsibility than normal, without the necessary staff support. As long as the operation was only a limited helicopter lift operation with a few wounded persons, the working load on this level was adequate. However, if the vessel had grounded and a mass evacuation had to be launched, the situation would have been different.

At the JRCC, all available SAR mission co-ordinators were mobilized, and roles distributed among them. Among other things, the logistics of bringing in new resources and also keeping up the capacities for other emergencies were important. This planning included activating the host nation support scheme from neighbouring countries.

At an operational level, alarming, mobilizing, and securing the logistics are important tasks at the start of an operation. For maritime operations, the RCCs take on the role of mobilizing the resources needed. In the *MV Viking Sky* case, extra helicopter capacity had to be mobilized from different sources, including Denmark and Sweden. More than the normal towing capacity for towing vessels was needed. The informational monitor and disseminator roles are also crucial, as the co-ordinator must continuously assess the risks of the situation and the personal safety of the involved actors on the scene. The RCC must continuously gather information, bring up satellite feeds and weather forecasts, and feed the information further on-scene.

There is a broad range of mission coordination tasks at RCCs. It is difficult for RCCs and company response organizations to uphold communication with the captain, especially when the bridge is no longer accessible. There is a need to plan communications between offshore rescue, air operations, and on-shore rescue. Authority was given from the duty officer to the other SAR mission co-ordinators present taking responsibility for co-ordinating the standby vessels, acquiring towing vessels, mobilizing air resources, and communicating with the local rescue co-ordination centre for shore operations.

We see that the special circumstances of the location of the the size of the operation and the duration of the action called for management with more specialized tasks and overlap between the co-ordinators to secure enough resources.

Informational Roles

The pilots on board the *MV Viking Sky* took the management role as information liaisons and advisers to the captain.

The captain of the distressed vessel must control information flow and be as accurate as possible. Too much information, especially if passengers provide information through smart devices, can lead to confusion and misinformation spreading. The captain must control information flow and be as accurate as possible. Therefore, informational roles could be beneficial for leaders and coordinators in the response operation. The case of *MV Viking Sky* demonstrated that in larger incidents there might arise challenges for informational roles on scene, both keeping the OSC and the SMC at operational level informed. In the *MV Viking Sky* accident, increased situational awareness was achieved through the liaisons from the Norwegian Coastal Administration. As a result, we experienced a new sort of structure with the NCA representatives taking an active managerial role in a SAR operation.

In large-scale complex operations, there is a need to manage a broad range of resources, operational agencies, and stakeholders. Government authorities and politicians, media, and the public will need a lot of follow-up. Other countries have to be contacted for additional resources using the Host Nation Support Guidelines. On board a large cruise vessel there could be 60 nations represented among the passengers and crew. Taking on the informational role at this level can call for tailoring of procedures in each case. There is a need to develop management roles with innovation capabilities. It is important to establish a network of response actors, industry and regulatory bodies and forums for information-sharing.

RCCs provide information to higher authorities such as a national governmental organization or ministry. The staff of these bodies must receive accurate information and training and skills in how to give information in a state of emergency. Also, national rescue boards could play an active role in both information and decision-making, taking care of more tasks to relieve the SMC at the operational level from some burdens. The information and co-ordination of the national level became a challenge as the DSB intervened and mobilized to co-ordinate between regional and national levels. However, this drew resources away from the co-ordination that the JRCC was doing.

Inter-personal Roles

Besides the informational roles, the captain of the vessel has important inter-personal roles and could be the key person to rearrange the roles of the crew. He has to take a figurehead role to avoid panic and calmly direct the safety crew. This approach would be of particular importance if the passengers on board had to be evacuated into lifeboats and liferafts.

Within the teams both at tactical and operational level, a long-term operation had to be prepared for with a lot of potential traumatic situations. The team leaders therefore had to be careful not to overstretch their personnel and mobilizing additional forces if necessary.

Decision-making Roles

In the *MV Northguider* case, the helicopter pilots together with the helicopter crew, in effect served as both OSCs ACOs on the scene. They had to rely on each other to search for the vessel, lift the crew from the *MV Northguider*, and also consider alternative paths for the operation. Also, owing to communication challenges and lack of co-ordinative support on scene, the police officer at the RSC in Longyearbyen in reality took over the role as SAR mission co-ordinator while the rescue operation took place. This did not follow the IAMSAR Manual, and the JRCC did not approve this in the report afterwards, noting that in case of rescue operations, RSC Svalbard should not co-ordinate resources without agreeing with the JRCC (Joint Rescue Co-ordination Centre, 2019).

Managerial Roles and Structuring Mechanisms

Structure Elaboration

We can see in the *MV Viking Sky* case that structure elaboration was important, especially on shore. At an operational level, within the police, the operational centre (112) capacity was increased. On board the vessel, the captain considered the evacuation of the passengers and crew. The conditions would have demanded a significant increase in the manning of each position to take care of especially elderly passengers, adding more personnel both from the crew and volunteers from the passenger list to take care of the evacuation and survival phase at sea. In polar regions with ice conditions, the Polar code recommends equipment for survival for five days before rescue. This planning will be very demanding for all persons evacuated and demand more co-operation between crew and passengers.

The case of the *MV Viking Sky* demonstrated the large number of organizations that the JRCC-SN alerted to get an overview of the available resources and create an emergency response value chain. Also, considering additional personnel because of potentially lengthy and complicated operations, structural elaboration can be pre-defined in co-ordinative organizations, as the key management personnel will be exhausted after a long day of operation. The *MV Viking Sky* case showed how important the proactive role of SMC planning is for the worst-case scenario. Within the volunteer organizations, several hundred persons were alerted and ready for action, should the operation escalate or go on for several days. Host nation support was also arranged from the neighbouring countries.

The RCC also has responsibilities to keep superiors informed. Liaising with embassies and government ministries and the EU-Emergency Response Co-ordination Centre would be important. This situation revealed the need for a new managerial role as an emergency response liaison officer from the shipping company or tour operators, and more links to the national political level.

The presence of the pilots on board *MV Viking Sky* serving as a liaison between the captain and the JRCC opens up a discussion on expanding the on-scene co-ordinator role with an OSC distress vessel. With SAR resources on board such as the medics, rescue men from the SAR-helicopter, fire-fighters, and police special forces, there would be a need for a broader on-scene co-ordination effort for a single OSC on board a Samaritan or coast guard vessel.

Role-switching

Distances between the units in distress and the emergency response units are of great concern in the Arctic. The first vessel on the scene might not be a professional resource like the coast guard or rescue vessels, but a vessel of opportunity. The distribution of workload between the SMC and the OSC would depend on the equipment, competence, and functioning of communication systems. If the OSC is not capable of taking on specific tasks, the SMC would have to put more staff and/or support on land. For masters of the vessels of opportunity, it is important to be ready for switching to this role and for the new structure elaboration when it comes to assigning the role of the OSC. Later on, authority migration and system resetting could be necessary when a professional SRU unit approaches the scene. Instead of being mostly in a decisional role, the OSC role pattern can be adjusted, depending on available resources, competence, and situation, by taking on a more informational role.

Role-switching was present as the pilots took the role as informants and liaisons between the captain and the JRCC in the *MV Northguider* case. The local OSC should, in particular, be prepared for role-switching. The Arctic weather could hamper some duties of the OSC. As in the case of the *MV Viking Sky*, the OSC was unable to help with the evacuation because of the high seas. The evacuation had to be performed by air; therefore the co-ordination of rescue resources was mainly taken care of by the ACO. On board the *MV Viking Sky*, role-shifting took place as the pilots took an informational role and also took part in the decision-making process serving the captain.

Authority Migration

In the *MV Northguider* case, the police officers at the County Governor's office took over the authority as temporary SAR mission co-ordinators after direction from the helicopter pilots. The helicopter pilots regarded the situation as so critical that they wanted to reduce the communication needs to a minimum, choosing the police operational centre at the County Governor as contact points. The County Governor confirmed the critical situation in the Evaluation report:

> Under the circumstances with the weather conditions on scene, the helicopter crew balanced close to the limits to save lives.
>
> Governor of Svalbard, 2019

This authority migration was not fully legitimized by the SAR mission co-ordinators, who made some recommendations afterwards to update and maintain the understanding of procedures about how the JRCC needs to co-ordinate missions at sea.

The inclusion of navigational pilots from the Coastal Administration in the vessel's SAR team was a migration of authority from the captain to the pilots.

In the case of the *MV Viking Sky*, the number of units related to helicopter rescue called for specialized expertise. The JRCC handled this by the delegation of roles such as the ACO to air traffic control to handle the large number of units involved. This case was not typical in maritime SAR operations.

Even though there are set structures in most countries for strategic-level SAR management, each major SAR operation such as a maritime evacuation and rescue action could call for a tailor-made solution. The *MV Viking Sky* case showed that the police were reluctant to push the 'big disaster alarm button'. Such a decision will take up a lot of limited resources to this specific action. This case shows that there is a huge responsibility in prioritizing resources, taking controversial decisions, and mobilizing a lot of costly government and private resources fast enough. Having the 'courage' at lower levels to influence the levels above could be challenging. The role-switching mechanisms between levels can be discussed in such fast-escalating cases as a major cruise ship accident.

System-resetting

In a situation where a major disaster could happen, the police staff and the local rescue board were not established despite the fact that this procedure could potentially be used. Instead, the police operational centre was given extra capacity under the chief of operations. The argument for this system-resetting was that the main issue was having an overview of the situation and controlling the units on the scene. The system-resetting is also a challenge on board the vessel. In a SAR operation, the shipowner is in charge of salvaging and taking care of the vessel, while the expedition operator is in charge of crew and passengers. The distress vessel captain will be the single authority as long as the vessel is afloat and the passengers and crew are on board. We can here find fluctuating tasks and roles as well as responsibilities. Task-sharing and a clear decision-making structure will be important particularly for providing information. There is a chain of authority and responsibility between the owner, operator, the distress vessel captain, and the emergency agencies, especially the SMC at the operational level. These circumstances call for flexibility in the discussion about how the system should work in different phases.

Conclusion

In this paper, we have illuminated the challenges of inter-organizational co-ordination in large-scale operations and the roles of the incident commanders at different levels when facilitating joint emergency response in complex operations. The results show that the range of managerial roles have to be reflected on and expanded to deal with the lack of predictability about emergency cause–effect relations, resources, and capabilities. The joint configuration of emergency organization includes multiple actors across many jurisdictions with diverging organizational design. In fast-escalating situations, there is a need for leaders taking over authority from other levels and agencies. Thus, there should be available role-switching mechanisms for critical situations in the commanding and co-ordinating organizations. These situations also call for special structuring mechanisms to co-ordinate multi-agency collaboration. The standardized ICS for co-operation and co-ordination of several organizations might not be sufficient to deal with high-complexity environments, owing to the need for flexible solutions, including system-resetting. Our findings show that the increased complexity demands managerial roles and mechanisms within the co-ordinating organizations that are adapted to the joint force configuration with context-related solutions. The lead organization could prepare for demanding changes in the organizational structures and managerial roles configuration for the key leaders through structuring mechanisms allowing for role-switching.

Further studies are needed to look into the balance between organizational standardization and structural and managerial flexibility. This research should focus on collecting in-depth data on the emergency response processes at different management levels and reflect on the range of roles and the related structuring mechanisms needed. Future research should also look closer into the range of joint teams and their co-ordination at different levels, comparing structuring mechanisms in different systems, with a particular focus on creativity and flexibility. The adaptation of managerial roles in different phases should be highlighted. Finally, future studies should look into how changes in tasks between management levels are handled over time, especially related to strategic co-ordination in long-term, large-scale response operations.

References

Andreassen, N., Borch, O.J., and Ikonen, E.S. (2018b). Managerial Roles & Structuring Mechanisms within Arctic Maritime Emergency Response. *The Arctic Yearbook 2018*: pp. 275–292. Retrieved from http://hdl.handle.net/11250/2591156.

Andreassen, N., Borch, O.J., and Ikonen, E.S. (2019). Organizing emergency response in the European Arctic: A comparative study of Norway, Russia, Iceland and Greenland. MARPART Project Report 5. Nord University, R&D-Report no. 46. Bodø.

Andreassen, N., Borch, O.J., Kuznetsova, S., and Markov, S. (2018c). Emergency Management in Maritime Mass Rescue Operations: The Case of the High Arctic, in L. Hildebrand et al. (Eds), *WMU Studies in Maritime Affairs, Vol. 7, Sustainable Shipping in a Changing Arctic*. Cham: Springer.

Andreassen, N., Borch, O.J., and Schmied, J. (eds.) (2018a.) Maritime emergency preparedness resources in the Arctic: capacity challenges and the benefits of cross-border cooperation between Norway, Russia, Iceland and Greenland. MARPART Project Report 4, R&D Report 33, Nord University. Retrieved from: http://hdl.handle.net/11250/2569868.

Bharosa, N., Lee, J.K., and Janssen, M. (2010). Challenges and obstacles in sharing and coordinating information during multi-agency disaster response: Propositions from field exercises. *Inf Sys Front*, 12: pp. 49–65.

Beck, T.E. and Plowman, D.A. (2014). Temporary, Emergent Interorganizational Collaboration in Unexpected Circumstances: A Study of the Columbia Space Shuttle Response Effort. *Organization Science*, 25(4): pp. 1234–1252.

Bigley, G. A. and Roberts, K.H. (2001.) The incident command system: high reliability organizing for complex and volatile environments. *Academy of Management Journal*, Vol. 44, no. 6: pp. 1281–1299.

Boin, A. and Hart, P. (2003). Public leadership in times of crisis: Mission impossible. *Public Administration Review*, 63: pp. 544–553.

Borch, O.J. (1994).. The process of relational contracting PRIVATE Developing Trust-Based Strategic Alliances among Small Business Enterprises. In P. Shrivastava, J. Dutton, and A. Huff (eds.), *Advances in Strategic Management*. Greenwich, CT: JAI Press Inc.

Borch, O.J. and Andreassen, N. (2015). Joint-Task Force Management in Cross-Border Emergency Response. Managerial Roles and Structuring Mechanisms in High Complexity-High Volatility Environments. In A. Weintrit and T. Neumann(eds.), *Information, Communication and Environment: Marine Navigation and Safety of Sea Transportation*. Boca Raton, FL: CRC Press.

Borch, O.J. and Batalden, B. (2014). Business-process management in high-turbulence environments: the case of the offshore service vessel industry. *Maritime Policy & Management*, Vol. 42, Issue 5: pp. 481–498.

Borch, O.J., Andreassen, N., Marchenko, N., Ingimundarson, V., Gunnarsdóttir, H., Jakobsen, U., Kern, B., Iudin, I., Petrov, S., Markov, S., and Kuznetsova, S. (2016a). *Maritime Activity and Risk Patterns in The High North*. MARPART Project Report 2. Bodø: Nord University: pp. 124. R&D report no. 4. Retrieved from: http://hdl.handle.net/11250/2432922.

Borch, O.J., Andreassen, N., Marchenko, N., Ingimundarson, V., Gunnarsdóttir, H., Iudin, I., Petrov, S., Jakobsen, U., and Dali, B. (2016b). *Maritime activity in the High North − current and estimated level up to 2025*. MARPART Project Report 1. Bodø: Nord University, (7): p. 130. Retrieved from: http://hdl.handle.net/11250/2413456.

Borch, O.J., Roud, E.P., Schmied, J., Berg, T.E., Fjørtoft, K., Selvik, Ø., Parsons, J, and Gorobtsov, A. (2016c). SARINOR WP7 Report -Behov for trening, øving og annen kompetanseutvikling innenfor søk- og redning i nordområdene. Retrieved from: www.sarinor.no/2016/08/16/sarinor-arbeidspakke-7/.

Buck, D.A., Trainor, J.E., and Aguirre, B.E. (2006). A Critical Evaluation of the Incident Command System and NIMS. *Journal of Homeland Security and Emergency Management*, Vol. 3, Issue 3, Article 1. Retrieved from: www.bepress.com/jhsem/vol3/iss3/1.

Campbell, D.J. (1988). Task complexity: A review and analysis. *Academy of Management Review*, 13: pp. 40–52.

Comfort, L.K. and Kapucu, N. (2006). Inter-organizational coordination in extreme events: The World Trade Center attacks, September 11, 2001. *Natural Hazards*, Vol. 39: pp. 309–327.

Cosgrave, J. (1996). Decision making in emergencies. *Disaster Prevention and Management: An International Journal*, Vol. 5, Issue 4, pp. 28–35.

Crichton, M.T., Lauche, K. and Flin, R. (2005). Incident Command Skills in the Management of an Oil Industry Drilling Incident: a Case Study. *Journal of Contingencies and Crisis Management*, 13(3).

Erdi, P.(2008). *Complexity Explained*. Berlin: Springer.

Flin, R., Slaven, G., and Stewart, K. 1996. Emergency Decision Making in the Offshore Oil and Gas Industry. *Human Factors*, 38(2): pp. 262–277.

Governor of Svalbard. (2019). Evaluation report. (Evaluering Etter Redningsaksjon i Hinlopenstredet), 28 December 2018.

High North News. (2019). Hadde «Viking Sky»-dramaet utspilt seg utenfor kysten av Nord-Norge, kunne konsekvensene blitt verre, 25 March. Retrieved from: www.highnorthnews.com/nb/hadde-viking-sky-dramaet-utspilt-seg-utenfor-kysten-av-nord-norge-kunne-konsekvensene-blitt-verre.

Hossain, L. and Uddin, S. (2012). Design patterns: coordination in complex and dynamic environments. *Disaster Prevention and Management*, Vol. 21, No. 3, pp. 336–350.

Hærem, T., Pentland, B.T., and Miller, K.D. (2015). Task complexity: extending a core concept. *Academy of Management Review*, Vol. 40, No. 3: pp. 446–460.

International Maritime Organization and International Civil Aviation Organization. (2016). *IAMSAR Manual, Volume I Organization and Management*. Tenth edition. London: IMO; Montreal: ICAO.

Bouty, I., Godé , C., Drucker-Godard, C., Lièvre , P., Nizet , J., and Pichault, F. (2012). Co-ordination practices in extreme situations. *European Management Journal*, Vol. 30: pp. 475–489.

Jardine-Smith, D. (2014). Mass rescue. *Seaways*, January.

Jensen, J. and Thompson, S. (2016). The incident command system: a literature review. *Disaster*. 40(1): pp. 158–182.

Joint Rescue Co-ordination Centre, Norway. (2019). Timeline Viking Sky. Retrieved from: www.hovedredningssentralen.no/tidslinje-viking-sky/.

Joint Rescue Co-ordination Centre, North Norway. (2019. Review report after the Northguider incident.

JRCCs Northern and Southern Norway. (2019). *SAR Co-operation Plan*, part III, IV, V and VI, February 2019. www.hovedredningssentralen.no/wp-content/uploads/2019/02/SAR-Cooperation-Plan-rev.pdf.

Kapucu, N. (2005). Interorganizational Coordination in Dynamic Context: Networks in Emergency Response Management. *Connections*, 26(2), pp. 33–48.

Klein, G. (2008). Naturalistic Decision Making. *Human Factors*, Vol. 50, No. 3, pp. 456–460.

Klein, G.A., Calderwood, R., and Clinton-Cirocco, A. (1986). Rapid decision making on the fireground. Proceedings of the Human Factors and Ergonomics Society 30th Annual Meeting, 1: pp. 576–580.

Marchenko, N., Andreassen, N., Borch, O.J., Kuznetsova, S., Ingimundarson, V., Jakobsen, and U. (2018). Arctic Shipping and Risks: Emergency Categories and Response Capacities, *International Journal on Marine Navigation and Safety of Sea Transportation*; Vol. 12, (1): pp. 107–114.

Mintzberg, H. (1973). *The Nature of Managerial Work*. New York: Harper Row.

Mintzberg, H. (2003). The manager's job: folklore and fact. In Reynolds, Henderson, Seden, Charlesworth, and Bullman (eds), *The Managing Care Reader*. Abingdon: Routledge.

Mintzberg, H. (2009). *Managing*. Williston, VT: Berrett-Koehler Publishers.

Moynihan, D.P. (2009). The Network Governance of Crisis Response: Case Studies of Incident Command Systems. *Journal of Public Administration Research and Theory*. Vol. 19, No. 4 (October): pp. 895–915.

National Research Council. (2006). *Facing Hazards and Disasters: Understanding Human Dimensions*. Washington, DC: The National Academies Press.

NRK. (2019). Den dramatiske evakueringa frå Viking Sky – dette har skjedd. 24 March 2019. Retrieved from: www.nrk.no/mr/viking-sky-fekk-trobbel-_-dette-har-skjedd-1.14487961.

Okhuysen, G.A. and Bechky, B.A. (2009). *Coordination in Organizations: An Integrative Perspective. Academy of Management Annals*, Vol. 3, no. 10: pp. 463–502.

Paton, D. and Flin, R. (1999). Disaster stress: an emergency management perspective. *Disaster Prevention and Management: An International Journal*, Vol. 8, Issue 4: pp. 261–267.

Phillips, N., Lawrence, T.B., Hardy, C. (2000). Inter-organizational collaboration and the dynamics of institutional fields. *Journal of Management Studies*, 37(1): 23–43.

Ranson, S., Hinings, B., and Greenwood, R.(1980). The Structuring of Organizational Structures. *Administrative Science Quarterly*, Vol. 25, No. 1: 1–17.

Singstad, A. and Dybfest, E. (2019). Dramaet i Hustadvika: '-det er sannsynlig at noe usannsynlig kommer til å skje'. Presentation at Exercise Nord, Nord University, Bodø, 23 April 2019.

Simon, H. (1957). *Models of Man: Social and Rational.* New York: Wiley.

The Barents Observer. (2019). The Viking Sky incident – A wake-up call for the Arctic cruise industry? 26 March 2019. Retrieved from: https://thebarentsobserver.com/en/travel/2019/03/viking-sky-incident-wake-call-arctic-cruise-industry.

Tierney, K. and Trainor, J. (2004). *Networks and resilience in the World Trade Center disaster. Research Progress and Accomplishments,* Vol. 6. Buffalo, NY: Multidisciplinary Center for Earthquake Engineering Research.

Turoff, M., White, C., and Plotnick, L. (2011). Dynamic Emergency Response management for large Scale Decision making in Extreme hazardous Events. In F. Burstein, P. Brezillon, and A. Zaslavsky (eds), *Supporting Real Time Decision-Making,* Vol. 13. New York: Springer.

USA Today. (2019). Viking Sky cruise timeline: A breakdown of what we know happened. 29 March 2019. Retrieved from: https://eu.usatoday.com/story/travel/cruises/2019/03/26/viking-sky-cruise-evacuation-and-rescue-timeline-what-happened/3275539002/.

Weick, K.E. and Sutcliffe, K.M. (2011). *Managing the unexpected: Resilient performance in an age of uncertainty.* London: John Wiley & Sons.

Wolbers, J., Boersma, K., and Groenewegen, P. (2017). Introducing a Fragmentation Perspective on Coordination in Crisis Management. *Organization Studies*: 1–26.

13 Counterterrorism at Sea and Its Implications for Selected Arctic Maritime Soft Targets

Jana Prochotska

Introduction

Crisis management and law enforcement activities within the Arctic maritime environment pose various challenges on parties involved, owing to strenuous natural conditions, vast areas, remote locations, and scarce resources (Elgsaas, 2018). Violent actions and terrorism with lives and several other negative consequences at stake are no exceptions. Even given their low probability nature in the geographical area examined, a non-anticipated incident could have unexpected impacts on stakeholders and the wider society. Places with limited or no security measures that gather many civilians − so-called soft targets − are among the most exposed. This makes, for example, cruise vessels, passenger ferries, and offshore drilling rigs potential targets of violent actions and terrorism (Greenberg et al., 2006). The US Department of Homeland Security describes soft targets and crowded places in their document 'Soft Targets and Crowded Places Security Plan Overview' (2018) as:

> Sports venues, shopping venues, schools, and transportation systems, are locations that are easily accessible to large numbers of people and that have limited security or protective measures in place making them vulnerable to attack.

Given its politicized, multi-faceted, and subjective nature, terrorism represents a highly contested concept, and neither academics nor practitioners have yet agreed on a universal definition. Furthermore, this phenomenon nearly always involves moral and subjective judgements, and the shift from being labelled a terrorist to be celebrated as a freedom fighter can happen over time (Myhten and Walklate, 2005). In this chapter, we apply the definition from the International Convention for the Suppression of the Financing of Terrorism (1999):

> Any ... act intended to cause death or serious bodily injury to a civilian, or any other person not taking part in hostilities in a situation of armed conflict, when the purpose of such acts, by its nature or context, is to intimidate a population or to compel a government or an international organization to do or to abstain from doing any act.

Subsequently, maritime terrorism is briefly defined as terrorism that takes place on board units or installations at sea. The Council for Security Co-operation in the Asia Pacific (Yau, 2017) refers to maritime terrorism as:

> The undertaking of terrorist acts and activities (1) within the maritime environment, (2) using or against vessels or fixed platforms at sea or in port, or against any of their passengers or personnel, (3) against coastal facilities or settlements, including tourist resorts, port areas and port towns or cities.

State terrorism and state-sponsored terrorism are not reflected upon in the chapter. In order to delineate the difference between traditional seaborne crimes and terrorism, the definitions of piracy and armed robbery are explored. Based on the 1982 United Nations Convention on the Law of the Sea (UNCLOS), piracy can be defined as:

> Any illegal acts of violence or detention, or any act of depredation, committed for private ends by the crew or the passengers of a private ship or a private aircraft, and directed (i) on the high seas, another ship or aircraft, or against persons or property on board such ship or aircraft; (ii) against a ship, aircraft, persons or property in a place outside the jurisdiction of any State; [...].

Resolution A.1025(26) (Annex, paragraph 2.2) of the International Maritime Organization's Code of Practice for the Investigation of Crimes of Piracy and Armed Robbery Against Ship, describes armed robbery against ships as follows:

> Any illegal act of violence or detention or any act of depredation, or threat thereof, other than an act of piracy, committed for private ends and directed against a ship or persons or property on board such a ship, within a State's internal waters, archipelagic waters and territorial sea; [...].

In simplified terms, what differentiates piracy from the armed robbery is the place where the act was executed; the act is considered piracy if performed on the high seas and outside the jurisdiction of any state, whereas armed robbery takes place within a state's internal waters, archipelagic waters, or territorial sea. Concepts of terrorism and piracy both represent violent acts at sea, and some of their tools overlap. Notwithstanding their similarities, there exist clear differences that are illustrated in the definitions mentioned above. In summary, pirates engage in criminal activities to achieve financial gains without conveying an ideological message to the broader audience and act in secrecy outside the media reach to avoid being detected by law enforcement authorities.

In contrast to piracy, terrorist groups aim to impose intimidation and fear onto the population in order to obtain their political, social, and ideological objectives and consider media crucial in broadcasting their cause. It is a piracy act against maritime security if persons are taken hostage to seek money from ransom, and no other motivation is involved. Terrorists also apply hostage-taking as a technique to gain financial profits, yet such criminal activity is used to support their propaganda campaign.

The objective of the chapter is to analyse the Arctic maritime environment as a possible target of terrorist attacks, with a focus on soft targets, and to discuss counterterrorism concepts responding to politically motivated violence. The analysis seeks to answer the following research questions:

- What are the main drivers behind potential terrorist activities in the Arctic marine domain? What role do unique features of selected Arctic maritime soft targets play in terrorists' decision-making?
- Which counterterrorism security measures are to be considered in connection with the selected Arctic maritime soft targets?

Theory

The Arctic can be selected as a locus of different terrorist activities. However, the chapter focuses specifically on maritime terrorism against soft targets. Theoretical discussion around target selection by terrorist organizations and countermeasures suggested within the Arctic nautical space is based on rational choice theory.

Figure 13.1 portrays the complex relationship structure between dependent and independent variables. Arctic terrorism is influenced by internal factors on the one hand (terrorist decision-making) and external factors on the other hand, which have an impact on terrorists' decision-making. Counterterrorism

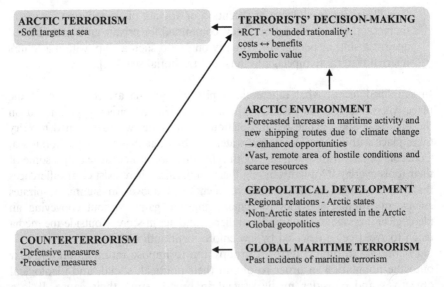

Figure 13.1 Analytical model

efforts normally reflect the security development, which in this case means addressing possible threats of terrorism, given the enhanced maritime activity and interest in the Arctic. Furthermore, the borderless nature of terrorism calls for regional and international co-operation to counter terrorist attacks, and such co-operation is shaped by geopolitical development. Counter-terrorism measures are additionally designed considering past terrorist incidents and expertise and lessons learned in connection with that. Terrorists' decision-making processes represent an influential factor for selecting functional counterterrorism concepts.

Rational Choice Theory and Target Selection

The rational choice approach consists of a tripartite model that attributes criminal behaviour to a minimum of three elements: a motivated offender, an opportunity to commit a crime, and the non-existence of a 'capable guardian' (Fussey, 2009). Underlying the rational choice theory, a decision-making process of an offender is explained by weighing the costs and benefits of an offence. Counterterrorism measures mainly respond to the components of an opportunity and guardianship by reducing possible target opportunities and improving the capability of safety measures. It is suggested that these actions will also manipulate the third element – motivation of an offender. In other words, it means that the higher the perceived risk of an offence, the lower the motivation of an individual to offend. Compared with other crimes, acts of terrorism and other politically motivated violence are influenced by values and normally carry a strong feeling of grievance. Given the absence of purely objective and value-free terms of politically motivated violence, there exist various difficulties in applying the interpretations of rational choice theory to these acts. The central concept of the rational choice theory of criminal activity is the objectivity of an offender's decision-making process. This assumption possesses practical relevance, but it is not exhaustive, owing to the exclusion of expressive and emotive factors and unpredictable components of human agency. Therefore, terrorist acts are explained within the notions of 'bounded rationality' (Fussey, 2009). The bounded rationality concept presented by Herbert Simon provides an alternative to the instrumental rationality discussed above. Within this concept, it is assumed that people are not capable of estimating all possible outcomes of their actions, but will make decisions based on their perception of the most crucial aspects of the consequence. Actors can still misjudge or ignore consequences.

Furthermore, bounded rational people do not examine all options available. The weighing of costs and benefits is stopped when an option, which is considered adequate and satisfactory, is found. The explanation lies in the high costs accrued by further searches, additional information being incomplete, erroneous, and possibly continuously changing. According to Simon's bounded rationality concept, a person does not maximize their utility but rather seeks an acceptable level of utility (van Um, 2009). Moreover, 'terrorists' decision-making' is formed by an influential factor, which is more prevalent in politically motivated violence than other types of criminal offences: the value of symbolism. In recognition of this, if a target possesses a symbolic value to a terrorist organization, such a target might be preferred even if the 'costs' are higher compared with another target type. For instance, anti-abortion terrorists will more likely attack a private clinic performing abortions than a general medical centre without the gynaecological department, even though the clinic has stronger security measures installed. The message conveyed by attacking a symbolic target is considered of higher 'benefit', and the target remains attractive regardless of the security system implemented, especially if there is no other symbolic target less guarded in the area of operation of the terrorist group.

When studying the target selection process and decision-making, it is necessary to reach beyond the notions of rational choice theory that considers the linkage between opportunity and motivation. In addition to the already discussed symbolism and grievance factors, research should also incorporate factors like the broader political context and social relations beyond the immediate surroundings (Fussey, 2019).

Counterterrorism

A discussion about the rationality of terrorists is vital in order to implement effective counterterrorism measures. If terrorists are recognized as instrumentally or bounded rational individuals, manipulating the utility function could mitigate the terrorism threat, whereas this strategy proves inefficient for the case of irrational terrorists. In light of cost-benefit calculations and suitable countermeasures, terrorists characterized by instrumental or bounded rationality are discouraged from committing violent acts by either increasing the costs of terrorism or decreasing the benefits of their acts. Costs can be increased by potential targets fortification or enhanced punishment. Benefits alteration seems more difficult to realize, as it normally requires major structural changes, which will lead to decentralization of political and economic power and thus lowering the impact of terrorist attacks (van Um, 2009). Increasing the costs of terrorist behaviour could lead to the displacement over time and space or into a different tactic (Fussey, 2019). For instance, the strengthening of borders of one target country could displace terrorist activity towards a country with the less secure territory (Sandler, 2015). The situational crime prevention (SCP) perspective claims that terrorism can be reduced by altering the opportunities that stand behind the success of terrorist acts (Freilich and LaFree, 2015). While SCP interventions prove valid in certain cases and there exists no doubt of their implementation necessity, terrorism researchers criticize the concept for its 'substitution effect', arguing that the measures do not provide for a decrease in terrorist events, but only shift the attacks against alternative targets, using different methods, other locations, etc. The introduction of metal detectors at US airports in 1973 reduced the number of skyjackings from an average of 25 to less than four per year (Sandler, 2015). Even if the regulation led to an increase of another terrorist method/type/location, the installation of metal detectors at airports has proved its high relevance over the years. Instead of avoiding and underestimating the function of SCP measures, one needs to remember the possible side effects of the measures. It needs to be acknowledged that SCP measures could transfer terrorist attacks from well-guarded hardened targets to more vulnerable places (Fussey, 2019). Places gathering a large number of people with limited or absent security measures in the Arctic are the main subject in the discussion in the next section of this chapter.

In his study, Sandler (2015) interprets counterterrorism as:

Actions to ameliorate the threat and consequences of terrorism. These actions can be taken by governments, military, alliances, international organizations (e.g. Interpol), private corporations, or private citizens.

There exist two main types of counterterrorism measures: defensive and proactive measures. Defensive countermeasures safeguard potential targets against terrorism by making attacks more costly to execute or reducing the probability of their success. Furthermore, when an attack has already emerged, the role of defensive measures is to mitigate the negative consequences of the attack. Defensive actions are commonly reactive, implemented after a certain development of terrorism. Examples of defensive or protective countermeasures involve target hardening (installing barriers, employing guards, technological surveillance, etc.), tougher penalties for terrorism offences, resilience-constructing, and improving capabilities of emergency and law enforcement actors.

Conversely, proactive or offensive countermeasures mean a confrontation of a targeted state with a terrorist group or its supporters. Proactive measures can target terrorist groups' training camps, financing structures, safe havens, propaganda websites, and their leaders and members. Other offensive measures include operations against state sponsors, infiltrating terrorist groups, military actions, and gathering intelligence to foil terror plots. Activities improving economic and social conditions and thus reducing grievances are also considered proactive (Sandler, 2015). Asymmetric threats like terrorism call for a multisectoral response and addressing root causes via linking security to development.

Methodology

This paper is explorative, underlying the lack of cases on the subject. The discussions are primarily centred around information and data from scientific articles, books, official and publicly available reports and publications of government institutions, international organizations, think-tanks, and the Global Terrorism Database. Secondary sources include media platforms, attending academic conferences, and off-the-record discussions with law enforcement and crisis management personnel.

The chapter employs qualitative and quantitative text analysis to study the novel research theme. Counterterrorism applied to the Arctic region represents a somewhat emerging and so far under-researched topic (Elgsaas, 2017), and the chapter aims to fill in some of the gaps, encourage continuous research and discussion, and point attention towards proactive threat assessments and the implementation of security measures.

Data and Analyses

Building upon the discussion in the Theory section, a starting point for the analysis is that the threat of maritime terrorism and counterterrorism measures in the Arctic need to be assessed and reflected upon in the context of three major factors:

- The changing shape of the Arctic environment in connection with climate change
- Global geopolitical developments with the focus on relations among the Arctic states
- Experience with maritime terrorism across the globe and application of expertise and lessons learned of counterterrorism programs concerning the Arctic particularities

Impact of Climate Change on the Arctic Environment

The 2014 assessment report of the Intergovernmental Panel on Climate Change pointed out that the Arctic has been warming at about twice the average global pace since 1980. The rising temperature is assumed to result in a nearly ice-free Arctic Ocean in summer periods and enhanced navigability of Arctic maritime space by 2050 (Connolly, 2017). The climate change can be translated into new economic opportunities as well as new security challenges. New shipping routes between Europe and Asia, namely the Northwest Passage and the Northern Sea Route, are consequences of the Arctic warming and declines in sea ice cover projected to it (Connolly, 2017). Positive benefits include, *inter alia*, shorter shipping routes[1], access to the extraction of natural resources, possible expansion of fisheries, new job opportunities, infrastructure development, knowledge generation, and tourism augmentation.

Conversely, the opening of the Arctic and an increase in maritime traffic could attract the operations of terrorist groups and other criminal syndicates that are interested in infiltration into the Arctic region and in use of Arctic trade routes for human, drug, and weapon trafficking (Nincic, 2012). Additionally, increased economic activity could increase the risks of oil pollution, accidents, environmental degradation, and interference in indigenous communities. This poses new requirements on response management, security infrastructure, intrastate partnerships, and interstate co-operation. Operations in the unique Arctic conditions of dark, ice, polar low-pressure systems, immense distances and scant resources call for specialist expertise and experience, state-of-the-art equipment and technology, multi-agency partnerships, and regional co-operation (Norwegian Shipowners' Association, 2014). It is worth noting that environmental development in the Arctic is not yet completely clear. Shorter routes and new opportunities for fossil extraction, fishing, and tourism are one aspect. However, the Arctic conditions will continue to challenge the costs of maritime operations, for instance in the form of necessary investment in suitable vessels or hiring icebreakers to accompany their vessels and higher insurance premiums.

Global Geopolitical Development and Relations Among Arctic States

Despite losing its strategic importance after the dissolution of the Soviet Union and the end of tensions between NATO and the Warsaw Bloc countries, the Arctic region has become once again a focal point, owing to the warming climate and re-emergence of geopolitical competition in the area (Connolly, 2017). Five of the Arctic states are members of NATO – Canada, Denmark, Iceland, Norway, and the USA. Denmark, Finland, and Sweden are part of the European Union. After the collapse of the Soviet Union, the co-operation of the Arctic countries focused on non-military security topics, such as environmental degradation and economic decline (Connolly, 2017). Currently, the priority of all Arctic states is to maintain the Arctic region as an area of stability, predictability, value-creation, responsible resource management, and peaceful co-operation, which is highlighted in the states' Arctic policy documents (see, for instance, Norway's Arctic Policy, 2014, and Sweden's Strategy for the Arctic Region, 2011). However, all five littoral states of the Arctic Ocean emphasize state sovereignty as one of the priorities. In legal terms, interstate relations are governed by the Arctic Council and the UNCLOS. On 1 January 2017 the mandatory International Maritime Organization Polar Code for ships operating in Arctic and Antarctic waters entered into force (Connolly, 2017). The UN's International Maritime Organization specializes in marine safety and environmental protection regulations concerning the global maritime industry, including the Arctic marine waters (Brigham, 2010). The diplomatic conduct of past maritime delimitation claims has confirmed the preference for peaceful co-operation within the Arctic region, in compliance with the maritime law. Research, marine oil pollution preparedness, aeronautical and maritime search and rescue emergency and preparedness demonstrate areas of intense co-operation of the Arctic states, to name but a few. The objective is to preserve the Arctic as a stable region of low levels of tensions, governance compliant with the maritime law, and peaceful co-operation (Norwegian Shipowners' Association, 2014).

The stability is reliant on the security posture and activities of the eight Arctic states, the conduit of non-Arctic countries pursuing their interest in being involved in the Arctic maritime space, and the presence of non-state illicit actors. In terms of both soft and hard security issues, the trust among the Arctic countries plays a vital role. Offensive rhetoric, power demonstrations, unpredictable behaviour, and international law violations do not create trustworthy partners. Terrorist groups operate in the borderless age of risks challenging the concept of national security, meaning that international counterterrorism efforts are imperative (Mythen and Walklate, 2005). While Arctic co-operation related to soft security benefits from an apolitical character and is well established in numerous areas, the same does not apply to hard security collaboration, which is prone to suspension as a result of geopolitical development across the globe.

Additionally, tagging a person or an organization as terrorist always involves politics (Elgsaas, 2019). Thus counterterrorism efforts in the Arctic can hardly reflect the tangible success of, for example, joint search and rescue operations and exercises. Arctic states' armed forces co-operation does exist to a certain extent, even beyond soft security matters. However, its further development is hindered by a lack of political trust within NATO–Russia relations. The missing confidence in an Arctic partner causes a significant hindrance for information-sharing, which is crucial for co-operative hard security operations. The Arctic Security Forces Roundtable represents a forum to discuss hard security matters in the Arctic, but Russia has not been invited since 2014 because of the Russian-Ukrainian conflict. Re-engaging Russia in the forum and encouraging debates on military transparency, incidents prevention, and regional conflicts management could demonstrate meaningful steps towards mutual trust rebuilding (van der Togt, 2019).

Global Maritime Terrorism

The Global Terrorism Database administered by the National Consortium for the Study of Terrorism and Responses to Terrorism (START) registered 357 terrorist incidents against maritime targets[2] in the period 1970 to 2017 (which accounts for 0.2% of all terrorist incidents) and 1,205 fatalities connected to it. The three deadliest sea terrorist attacks in this period are represented by an attack against a passenger steamer in Sudan by Christian extremists (13 February 1984), causing some 300 casualties; an attack against a ferry in the Philippines by Abu Sayyaf Group (27 February 2004), causing 116 casualties; and an attack against a boat carrying civilians in Yemen by Houthi extremists (Ansar Allah) (06 May 2015), causing 86 casualties. In the eight Arctic states, the database includes seven maritime terrorist attacks against the USA (Florida, Puerto Rico, New Jersey, and Michigan) and one attack against Denmark (Central Jutland). However, none of these attacks took place in the Arctic environment.

The number and consequences of violent acts at sea worldwide have proved that the maritime domain is significantly vulnerable and the absence or inadequacy of effective regulations and measures often lead to an increase of illicit activities that result in deterioration of marine environment, economic development, social stability, and national and human security (Curran, 2019). Porous coastlines, weak control mechanisms and capabilities, insufficient infrastructure, opportunity redundancy, and cases of corrupt personnel attract all types of complex criminal activities such as sea piracy, non-state terrorism, organized crime, etc. The maritime environment provides violent non-state actors with an alternative to circumvent strict onshore regulations and control measures.

Table 13.1 illuminates possible activities within the maritime environment orchestrated by terrorist organizations. The activities are logically organized in three main categories (inspired by Curran, 2019).

Table 13.1 Terrorist activities in the maritime domain

1. TACTICAL AND OPERATIONAL SUPPORT

a. Raids by sea on land targets
The sea is used to avoid onshore regulations, checkpoints, and border crossings. An attack is launched from a vessel, and weak coastal surveillance and control infrastructure enable perpetrators to stay undetected and proceed towards targeting the land objects.

b. Movement of fighters/weapons
Similarly to raids by sea, terrorist groups take advantage of low maritime space awareness, insufficient port security, and lack of resources and capabilities to move their fighters and weapons by circumventing overland security measures.

2. MARITIME DOMAIN AS A TARGET

a. Maritime terrorism
Terrorism at sea can take many variations. In addition to the financial gains explained below, the tactics include hostage-taking, ship hijacking, bombing ports, sea infrastructure, offshore platforms, sabotage, extortion plots, attacks against passenger ships and ferries, and others. Ships are used as floating bombs, as a weapon designed to hit the target or for launching explosives. Small boats with suicide bombers and remote-controlled vessels belong to operations employed by some terrorist groups.

b. Cyber-terrorism
Shipping industry with numerous ships at sea on any day and highly dependent on navigational and logistical systems is prone to cyber-attacks. Operation disruption may cause substantial economic losses given the present just-in-time concept of the international shipping system. Cyber-attacks may target ports, sea installations, and undersea fibre-optic cable network. Property damage and environmental degradation are not the only negative consequences possible. Illicit infiltration into shipping navigational systems can result in ship crashes with mass casualties.

3. FINANCIAL GAINS TO SUPPORT THE CAMPAIGN

a. Kidnap for ransom
Kidnapping passengers for ransom demonstrates another opportunity for terrorist groups to finance their operations. Cruise ships gathering wealthy travellers and high-value cargo vessels improve the probability of larger ransoms to be paid.

(continued)

Table 13.1 (continued)

b. Oil bunkering
Oil bunkering means stealing or adulterating of fuel products from legitimate oil and gas companies. Two-thirds of global oil exports are transported by sea. Accessing a high-pressure pipeline and diverting the flow of oil is the most frequently utilized method and largely undetectable, as it is commonly performed underwater, leaving the pipeline completely functional. With fuel products worth US$130 billion being stolen each year (Curran, 2019), this activity robustly contributes to global security destabilization.

c. Robbery
Increase in commercial maritime traffic leads to enhanced opportunities. Much of the sea trade passes through narrow and congested chokepoints where the vessels are obliged to slow down for safety reasons and thus become more exposed to felonious groups.

d. Drug/weapons-trafficking
Containerized shipping system has boosted international trade while enabling an illicit organization to easier access to drug and human trafficking. Given the enormous amount, only a fraction of these shipments can be inspected. This situation is particularly dangerous if the smuggling routes are used to move fighters and weapons of mass destruction.

e. Maritime migration
This activity comprises of migrant smuggling and migrant trafficking. Migrant smuggling means the procurement, in order to obtain a financial or other material benefits, of the illegal entry of a person into a state of which the person is not a national or permanent resident (Protocol against the Smuggling of Migrants by Land, Sea and Air, 2000). Human trafficking refers to the process through which individuals are trafficked to be placed or maintained in an exploitative situation for economic gain. Exploitative purposes include forced labour in factories, farms and private households, sexual exploitation, forced marriage, and forced involvement in terrorist organizations (Office of the United Nations High Commissioner for Human Rights, 2014).

f. Money-laundering
Licit enterprises involved in, for instance, fishing and oil and gas industries can be used by terrorist organizations for money-laundering or to conceal shipments of drugs, money, and weapons.

g. Taxation and extortion
Terrorist groups forcibly collect funds from the communities where they exert control over a territory. Illegal taxation, extortion, confiscation and looting account for 26% of total illicit flows to armed groups in conflict (Global Initiative against Transnational Organized Crime, 2018).

Soft Targets

From 1970 to 2017 the Global Terrorism Database registered nine terrorist attacks[3] against maritime targets with more than 20 people killed in each attack. Six incidents targeted civilian maritime vessels – ferries and boats carrying passengers. Five incidents involved bombing/explosion, and one attack took the form of an armed assault. The soft maritime target discussion in the chapter concentrates on cruise ships and passenger ferries as they assemble substantial numbers of people within a single physical location and lack satisfactory defence mechanisms. Three scenarios are put forward to elaborate further on the costs and benefits discussion of selecting targets by instrumental or bounded rational terrorists:

1   An attack against a cruise ship/passenger ferry via bombing/explosion or armed assault from land, another vessel, air transport, or on board of the cruise ship/passenger ferry, or using another vessel/air transport as a floating/flying bomb
2   Food/water poisoning on board of the cruise ship/passenger ferry
3   Taking passengers hostage for financial gains or negotiation purposes

Costs include:

- Logistics
- Procurement of transport means and weapons
- Planning
- Recruitment
- Training
- Incarceration, injury, or death
- Loss of reputation, loss of sympathizers, financial loss
- Detection

Benefits can comprise:

- Maintenance or change of the status quo for which the perpetrators engage in terrorist activities
- Financial gains
- Broadcasting their ideological message
- Attracting new sympathizers/members
- Lowering the government's credibility in the eyes of their population
- Injuries and fatalities
- Psychological suffering, fear
- Business disruption

It is widely held that terrorists resort to launching an attack if the benefits outweigh the costs (Hausken, 2016). Terrorists are not currently 'visible' around the Arctic coastal regions, and their operations in the Arctic environment would require a specific set of mariner skills and capabilities, as well as suitable transport means. Moreover, terrorist organizations prefer to adhere to methods and targets that they have already tried and mastered, providing higher chances of success and predictability. Owing to the offshore location of a target, an incident stays out of immediate reach of media and thus does not offer the same level of publicity as an attack on land-based objects (Chalk, 2008). Past sea terrorist incidents, however, show that the maritime domain is not immune to terrorism, and the Arctic region certainly represents a strategic opportunity.

Successful maritime terrorism across the globe could encourage copycat attacks in the Arctic. Factors contributing to the attractiveness of Arctic passenger vessels for terrorist organizations include publicly available schedules for easier pre-attack planning (the itineraries can be adjusted, though based on the actual weather conditions), a large volume of people within a single physical space, remote distance from onshore defence and emergency capabilities, certain vehicles being susceptible to capsizing (e.g. vehicle ferries), relatively low procurement expenses in relation with maritime improvised explosive devices, and 'low tech' possibilities, such as ramming a vessel into the target, etc. What serves to the advantage of terrorists is the open and public character of soft targets. Harder security measures could be socially undesirable and interfere with business practices. Integration of a 'security by design' concept from an early stage of soft targets development is highlighted in many soft targets protection guidelines (see, for instance, Action Plan to Support the Protection of Public Spaces, 2017).

Marine-based tourism represents the largest segment of the Arctic tourism industry in terms of numbers of persons, as well as geographical range (Arctic Marine Shipping Assessment, 2009), providing for an extensive number of opportunities and potential human and economic losses. Symbolic value embedded in attacks against cruise ships and passenger ferries in the Arctic can be explained as attacking the 'Western' lifestyle, political statements, and conveying a message that terrorist organizations are capable and willing to strike anywhere in the world. High-profile maritime attacks against large numbers of civilians can shake the credibility of liberal democracies. Seeking the disruption of relations between the government and their population, terrorist organizations specifically select targets that will communicate the political setting's incompetence to protect the lives of citizens (McCartan, Masselli, Rey, and Rusnak, 2008). Aiming to cause mass casualties of innocent people could, however, jeopardize the cause of terrorist groups.

Arctic Counterterrorism

Applied to the Arctic environment, counterterrorism is exposed to the same conventional challenges as management of other crises in the Arctic, such as oil spill, fire, and search and rescue operations. The challenges include harsh weather conditions (such as ice, icebergs, icing, reduced visibility), long distances, lack of infrastructure, and scarce resources (Borch et al., 2016). The security measures must be capable of recognizing threats, reducing the vulnerability of possible targets, mitigating the negative consequences, and ameliorate co-operation.

It needs to be emphasized that security at sea does not end at sea. Land-based security measures have a significant impact on maritime security, and onshore SCP efforts for cruise ships and passenger ferries include coast monitoring, port security management, screening of luggage to be boarded on a vessel, restricted areas, personnel background checks, and passenger screening. The significance of preventive regulations is well demonstrated by Abu Sayyaf's terrorist attack against SuperFerry 14 in the Philippines in 2004, during which the terrorists detonated explosives stored in an emptied out television set. A fire broke out as a result of the detonation and caused 116 fatalities (Yau, 2017).

SCP measures on board of a passenger vessel consist of, *inter alia*, security officers, ship security alert systems, contingency plans, emergency resources, training and drills, logs/records, and CCTV. As for soft targets, in particular, the security system must aim for the balance between target hardening and its natural open and public character (Action Plan to Support the Protection of Public Spaces, 2017). Different stakeholders possess different sets of functions usable and relevant before and during terrorist attacks. The well-informed general public can not only swiftly report suspicious behaviour, but, as they are often among first responders, they can also influence the course of an attack by disarming a perpetrator, requesting assistance, or helping to hide children. Creating public awareness and teaching the general public how to react in life-threatening situations supports a more resilient society development.

Operators of cruise ships and passenger ferries enhance the security on board by employing security contractors and benefiting from the use of specialized training for the ship personnel. Governments must encourage military-civilian and public-private co-operation to leverage the total resource and knowledge base. Terrorist attacks against cruise liners and passenger ferries could involve parallel mass rescue, search and rescue, fire, and oil spill response operations (Andreassen, Borch, and Ikonen, 2019). Contingency plans must be in place for all the scenarios. Broader regional co-operation needs to be established among the Arctic states to secure potent awareness and intelligence-sharing. With regards to the above-elaborated debate on relations between the Arctic states, co-operative efforts against terrorism within the setting of all eight Arctic states could be difficult to achieve. Partnerships with fewer Arctic countries should not be underrated, and the co-operation can take various forms – fusion centres, forums to exchange best practices and guidelines, military training, provision of escort vessels for high-capacity cruise liners, and joint patrolling and surveillance tasks. Initiatives encouraging regional maritime situational awareness serve as an important countermeasure. For instance, the information-sharing platform, Mercury, facilitates international efforts against piracy off the coast of Somalia by enabling different actors to feed and access real-time data and to communicate with each other (Bueger and Edmunds, 2017).

The inspiration for a comprehensive approach in terms of regional security improvement can be found in regions already dealing with sea terrorism incidents, such as the collaboration of ASEAN members in the fight against terrorism through confidence-building, intelligence-sharing, capacity-building and interoperability strengthening (Yau, 2017). The Nordic countries –

Denmark, Finland, Iceland, Norway, and Sweden – profit from sub-regional foreign and security policy co-operation, which is characterized by long tradition, shared values, mutual trust, high levels of public support for Nordic partnerships, flexibility, pragmatism, and informality. The co-operation is non-committal, and the primary attention is placed on information-sharing, exchange of perspectives, and joint analysis of foreign and security policy matters (Iso-Markku, Innola, and Tiilikainen, 2018).

It needs to be emphasized that no single counterterrorism strategy can function as an optimal response to prevent or mitigate all types of terrorist attacks and protect all sorts of targets. Counterterrorism measures ought to be tailor-made to reflect a particular type, motivation, modus operandi, and potential targets of a terrorist group (van Um, 2019).

Conclusion

The Arctic is currently undergoing a rapid transition and is becoming once again a strategic location and generator of new opportunities. Whereas economic prospects and environmental development can be relatively well estimated and acted upon sustainably, security threats such as politically motivated violence remain heavily unpredictable. Thus, this subject has to be re-assessed continuously to reflect upon further developments in the area. This chapter has strived to provide the contextual considerations, theoretical background, and practical suggestions to be included in counterterrorism debates fitting the peculiarities of the Arctic region. Despite the plethora of terrorism possibilities, the chapter focuses on potential soft targets. The analysis explains the aspects of terrorists' decision-making regarding cruise ships and passenger ferries operating in the Arctic. We argue that enhanced target opportunities resulting from environmental changes in the region together with symbolic value, advantageous cost-benefit calculation, the possibility to cause mass casualties and past incidents, could serve as a pull factor for terrorist organizations. Several measures are taken by the passenger vessel industry to reduce the risk. Our counterterrorism discussion reflects upon defensive and proactive security measures to be considered for implementation, and the challenges to be addressed. We acknowledge the necessity to understand the vulnerability and specifics of soft targets, as well as environmental conditions when designing a counterterrorism concept. Furthermore, wider regional co-operation within hard security issues is assumed to be imperative, yet difficult to obtain without mutual trust. Nonetheless, platforms gathering certain Arctic countries for sub-regional counterterrorism co-operation have an important function.

Limitations and Implications

Limitations

It is worth noting that the effectiveness of counterterrorism measures is scarcely discussed in the scientific literature (van Um, 2009). Evaluation of such measures most probably exists in reports and documents of practitioners resulting from actual incidents or exercises, but these assessments are of a classified nature. As there have not been any terrorist attacks in the Arctic region, the data on the presence, activities, and foiled plots of terrorist organizations represent a fruitful source for analysis. However, once again, this information is not publicly available. Additionally, terrorism research relies on much scarcer incidents than other criminological research, making the analysis leading to patterns and trends generation more challenging (Lum, Kennedy, and Sherley, 2009). Connecting implemented security measures with evaluations of their impacts is central to efficient counterterrorism strategies.

Implications for Further Research and Practical Implications

The results indicate the importance of additional research on the effectiveness of counterterrorism measures, possibilities for active regional co-operation in hard security matters, and identification of roles and gaps of the stakeholders involved in the response to and prevention of terrorist attacks. Our study encourages proactive multisector and cross-national co-operation in terms of threat assessment and consequent implementation and adjustment of security measures capable of prevention or mitigation of terrorist attacks against Arctic maritime soft targets. The aviation industry, as well as regions that have experience of maritime terrorism, represent a fruitful source of expertise.

Notes

1 Northwest Passage decreases the route between the Atlantic and the Pacific by 8,000 km and the Northern Sea Route between the Pacific and Western Europe by 5,000 km (Buky, 2009).
2 Maritime targets within the Global Terrorism Database include ports and maritime facilities. Maritime targets represent civilian maritime such as attacks against fishing ships, oil tankers, ferries, and yachts. Attacks on individual fishermen are excluded from this category, as they are coded in the database as 'Private Citizens and Property'.
3 There are 52 terrorist attacks against maritime targets registered in the Global Terrorism Database without a known number of casualties.

References

Akers, R.L. (1990–1991). Rational Choice, Deterrence, and Social Learning Theory in Criminology: The Path Not Taken. *Journal of Criminal Law and Criminology*, 81(3): 653–676.

Aktar, N. (2018). Terrorism at Sea: Role of International Legal Instruments and Challenges Ahead. *Research and Analysis Journal*, 1(1): 12–22.

Andreassen, N., Borch, O.J., and Ikonen, E. (2019). Organizing Emergency Response in the European Arctic: A Comparative Study of Norway, Russia, Iceland and Greenland, MARPART Project Report 5. NORD University, Norway. Arctic Council. (2009). Arctic Marine Shipping Assessment 2009 Report (second ed.).

Retrieved from: https://pame.is/images/03_Projects/AMSA/AMSA_2009_report/AMSA_2009_Report_2nd_print.pdf.

Asal, V.H., et al. (2009). The Softest of Targets: A Study on Terrorist Target Selection. *Journal of Applied Security Research*, 4(3): 258–278.

Borch, O.J. et al. (2016). Maritime Activity in the High North - Current and Estimated Level up to 2025, MARPART Project Report 1. NORD University, Norway.

Brigham, L.W. (2010). The Fast Changing Maritime Arctic: Globalization, Climate Change, and Geopolitics Converge in This Already Challenging Region. *Proceedings*, 136(5): 54–59.

Buky, M. (2009). Terrorism, Piracy and Climate Change: Challenges to International Maritime Governance. *Social Alternatives*, 28(2): 13–17.

Canetii-Nisim, D., Mech, G., and Pedahzur, A. (2007). Victimization from Terrorist Attacks: Randomness or Routine Activities? *Terrorism and Political Violence*, 18(4): 485–501.

Chalk, P. (2008). *The Maritime Dimension of International Security: Terrorism, Piracy, and Challenges for the United States*. RAND Project Air Force. Retrieved from: www.rand.org.

Clarke, R.V. (1997). *Situational Crime Prevention: Successful Case Studies (Second Edition)*. New York: Harrow and Heston Publishers.

Connolly, G.E. (2017). *NATO and Security in the Arctic (Report)*. NATO Parliamentary Assembly, Political Committee. Retrieved from: www.nato-pa.int/document/2017-nato-and-security-arctic-connolly-report-172-pctr-17-e-rev1-fin.

Curran, M. (2019). *Soft Targets & Black Markets: Terrorist Activities in the Maritime Domain*. One Earth Future. doi:10.18289/OEF.2019.038.

Dalaklis, D., Baxevani, E., and Siousiouras, P. (2018). The Future of Arctic Shipping Business and the Positive Influence of the International Code for Ships Operating in Polar Waters. *Journal of Ocean Technology*, 13(4): 78–94.

Dams, T. and van Schaik, L. (2019). *The Arctic Elephant: Why Europe Must Address the Geopolitics of the High North*. Policy Brief, Clingendael Netherlands Institute of International Relations. Retrieved from: www.clingendael.org/publication/arctic-elephant-europe-geopolitics-high-north.

Deflem, M.. (2006). Europol and the Policing of International Terrorism: Counter-Terrorism in a Global Perspective. *Justice Quarterly*, 23(3): 336–359.

Dodds, K. (2010). Flag Planting and Finger Pointing: The Law of the Sea, the Arctic and the Political Geographies of the Outer Continental Shelf. *Political Geography*, 29(2): 63–73.

Edmunds, T., Bueger, C.. (2017). Beyond Seablindness: A New Agenda for Maritime Security Studies. *International Affairs*, 93(6): 1293–1311.

Elgsaas, I.M. (2019). Arctic Counterterrorism: Can Arctic Cooperation Overcome its Most Divisive Challenge Yet? *The Polar Journal*, 9(1): 27–44.

Elgsaas, I.M. (2018). The Arctic in Russia's Emergency Preparedness System. *Arctic Review on Law and Politics*, 9: 287–311.

Elgsaas, I.M. (2017). Counterterrorism in the Russian Arctic: Legal Framework and Central Actors. *Arctic and North*, 29: 110–132.

European Commission. (2017). Action Plan to Support the Protection of Public Spaces. COM(2017) 612 final of 18.10.2017. Retrieved from: https://ec.europa.eu/home-affairs/sites/homeaffairs/files/what-we-do/policies/european-agenda-security/20171018_action_plan_to_improve_the_protection_of_public_spaces_en.pdf.

Farrell, G. (2010). Situational Crime Prevention and Its Discontents: Rational Choice and Harm Reduction versus 'Cultural Criminology. *Social Policy and Administration*, 44(1): 40–66.

Freilich, J., LaFree, G. (2015). Criminology Theory and Terrorism: Introduction to the Special Issue. *Terrorism and Political Violence*, 27(1): 1–8.

Fussey, P. (2009). An Economy of Choice? Terrorist Decision-Making and Criminological Rational Choice Theories Reconsidered. *Security Journal*, 24(1):1–15.

Global Initiative against Transnational Organized Crime. (2018). World Atlas of Illicit Flows. Retrieved from: https://globalinitiative.net/world-atlas-of-illicit-flows/.

Greenberg, M.D. et al. (2006). Maritime Terrorism: Risk and Liability. RAND Centre for Terrorism Risk Management Policy. Retrieved from: www.rand.org/pubs/mono graphs/MG520.html.

Hausken, K. (2016). A Cost-Benefit Analysis of Terrorist Attacks. *Defence and Peace Economics*, 29(2): 111–129.

Heath-Kelly, C. (2012). Counter-Terrorism and the Counterfactual: Producing the 'Radicalisation' Discourse and the UK PREVENT Strategy. *The British Journal of Politics and International Relations*, 15(3): 394–415.

International Maritime Organization. (2009). Resolution A.1025(26) on IMO's Code of Practice for the Investigation of Crimes of Piracy and Armed Robbery Against Ship. Retrieved from: www.imo.org/en/KnowledgeCentre/IndexofIMOResolutions/Assembly/Documents/A.1025(26).pdf.

Iso-Markku, T., Innola, E., and Tiilikainen, T. (2018). A Stronger North? Nordic Cooperation in Foreign and Security Policy in a New Security Environment. Government's Analysis, Assessment and Research Activities, Prime Minister's Office. Retrieved from: www.nordefco.org/files/nordic-vnteas-report_final.pdf.

Jesus, J.L. 2003. Protection of Foreign Ships against Piracy and Terrorism at Sea: Legal Aspects. *The International Journal of Marine and Coastal Law*, 18(3): 363–400.

Kalvach, Z. et al. (2016). Basics of Soft Targets Protection – Guidelines (2nd ed.). Soft Targets Protection Institute.

Karlos, V., Larcher, M., and Solomos, G. (2018). Review on Soft Target/Public Space Protection Guidance (2nd ed.). Joint Research Center Science for Policy Report. Retrieved from: http://publications.jrc.ec.europa.eu/repository/bitstream/JRC110885/soft_target-public_space_protection_guidance.pdf.

Lackenbauer, W., Huebert, R., and Dean, R. (2017). *(Re)Conceptualizing Arctic Security. Selected Articles from the Journal of Military and Security Studies*. University of Calgary. Retrieved from: https://cmss.ucalgary.ca/sites/cmss.ucalgary.ca/files/pwl-rd-rh-reconceptualizing-arctic-security-2017.pdf.

Lanteigne, M. (2019). The Changing Shape of Arctic Security. *NATO Review*. Retrieved from: www.nato.int/docu/review/2019/Also-in-2019/the-changing-shape-of-arctic-security/EN/index.htm.

Lum, C., Kennedy, L.W., and Sherley, A.J. (2009). The Effectiveness of Counter-Terrorism Strategies. *Campbell Systematic Reviews*. The Campbell Collaboration. Retrieved from: https://onlinelibrary.wiley.com/doi/pdf/10.4073/csr.2006.2.

McCartan, L.M., Masselli, A., Rey, M., and Rusnak, D. (2008). The Logic of Terrorist Target Choice: An Examination of Chechen Rebel Bombings from 1997–2003. *Studies in Conflict & Terrorism*, 31(1): 60–79.

Ministry of Foreign Affairs, Norway. (2014a). Security in the Arctic – a Norwegian Perspective. Speech by the State Secretary of the Ministry of Foreign Affairs Bard

G. Pedersen during the Arctic Circle conference in Iceland, 2 November. Retrieved from: www.regjeringen.no/en7dep7ud7id833/.

Ministry of Foreign Affairs, Norway. (2014b). Norway's Arctic Policy: Creating Value, Managing Resources, Confronting Climate Change and Fostering Knowledge. Developments in the Arctic Concern Us All. Retrieved from: www.regjeringen.no/globalassets/departementene/ud/vedlegg/nord/nordkloden_en.pdf.

Ministry of Foreign Affairs, Sweden. (2011). *Sweden's Strategy for the Arctic Region.* Retrieved from: www.government.se/49b746/contentassets/85de9103bbbe4373 b55eddd7f71608da/swedens-strategy-for-the-arctic-region.

Mythen, G., Walklate, S. (2005). Criminology and Terrorism: Which Thesis? Risk Society or Governmentality? *The British Journal of Criminology,* 46(3): 379–398.

National Consortium for the Study of Terrorism and Responses to Terrorism (START). (2018). *Global Terrorism Database [Data file].* Retrieved from: www.start. umd.edu/gtd.

Nincic, D.J. (2012). Maritime Security in the Arctic: The Threat from Non-state Actors. The 13th Annual General Assembly of the IAMU. Retrieved from: http://iamu-edu.org/wp-content/uploads/2014/07/Maritime-Security-in-the-Arctic-The-threat-from-non-state-actors.pdf.

Norwegian Shipowners' Association. (2014). High North – High Stakes: Maritime Opportunities in the Arctic. Retrieved from: https://rederi.no/en/rapporter/.

Office of the United Nations High Commissioner for Human Rights. (2014). Human Rights and Human Trafficking (Fact Sheet No. 36). Retrieved from: www.ohchr. org/Documents/Publications/FS36_en.pdf.

Peter, C.P. (2016). Cruising with Terrorists: Qualitative Study of Consumer Perspectives. *International Journal of Safety and Security in Tourism/Hospitality,* 14: 1–12.

Sandler, T. (2015). Terrorism and Counterterrorism: An Overview. *Oxford Economic Papers,* 67(1): 1–20.

Schopmans, H. (2019). Revisiting the Polar Code: Where Do We Stand? The Arctic Institute, Center for Circumpolar Security Studies. Retrieved from: www.thearcti cinstitute.org/revisiting-polar-code/.

Singh, A. (2019). Maritime Terrorism in Asia: An Assessment. ORF Occasional Paper No. 215, Observer Research Foundation. Retrieved from: www.orfonline.org/wp-content/uploads/2019/10/ORF_OccasionalPaper_215_MaritimeTerrorism-Asia. pdf.

Staalesen, A. (2015). Terrorism Is Coming to Arctic, Putin Fears. *The Barents Observer.* Retrieved from: https://thebarentsobserver.com/ru/node/254.

Stampnitzky, L. (2011). Disciplining an Unruly Field: Terrorism Experts and Theories of Scientific/Intellectual Production. *Qualitative Sociology,* 34: 1–19.

Stokke, O.S. (2011). Environmental Security in the Arctic: The Case for Multilevel Governance. *International Journal,* 64(4): 835–848.

Titley, D.W., St. John, C.C. (2010). Arctic Security Considerations and the U.S. Navy's Roadmap for the Arctic. *Naval War College Review,* 63(2): 35–48.

Tuerk, H. (2008). Combating Terrorism at Sea – The Suppression of Unlawful Acts against the Safety of Maritime Navigation. *University of Miami International and Comparative Law Review,* 15(3): 337–367.

United Nations. (1982). Convention on the Law of the Sea. Retrieved from: www.un. org/depts/los/convention_agreements/texts/unclos/unclos_e.pdf.

United Nations. (2000). Protocol against the Smuggling of Migrants by Land, Sea and Air, Supplementing the United Nations Convention against Transnational

Organized Crime. Retrieved from: www.unodc.org/documents/middleeastand northafrica/smuggling-migrants/SoM_Protocol_English.pdf.

US Department of Homeland Security. (2018). Soft Targets and Crowded Places Security Plan Overview. Retrieved from: www.hsdl.org/?collection&id=86643.

van der Togt, T. (2019). Conflict Prevention and Regional Cooperation in the Arctic. *Clingendael Magazine*. Retrieved from: www.clingendael.org/publication/conflict-prevention-and-regional-cooperation-arctic.

van Um, E. (2009). Discussing Concepts of Terrorist Rationality: Implications for Counter-Terrorism Policy. Economics of Security Working Paper 22. Deutsches Institut fur Wirtschaftsforschung, Berlin.

Yau, J.M.W. (2017). Maritime Terrorism Threat in Southeast Asia and Its Challenges. *Pointer, Journal of the Singapore Armed Forces*, 43(2): 32–44.

14 Emergency Collaboration Exercises and Learning

Experiences from the Arctic

Ensieh Roud and Johannes Schmied

Introduction

A large-scale emergency response typically requires co-ordinated action between multiple actors across many jurisdictions (Kapucu, Arslan, and Demiroz, 2010). Private, public, and volunteer organizations unite competencies and resources to respond to and resolve complex situations (Berlin and Carlström, 2011). Therefore, collaboration among the different organizations is essential to ensuring an effective emergency response. However, collaboration is often problematic in practice (Chen, Sharman, Rao, and Upadhyaya, 2008). Collaboration calls for better preparation through training in general and exercises in particular (Kristiansen, Løwe Sørensen, Carlström, and Inge Magnussen, 2017; Roud, Borch, Jakobsen, and Marchenko, 2016). Specifically, emergency collaboration exercises (ECEs) are designed to develop and test cross-sectoral and inter-organizational collaboration, preparedness efforts, and response quality in joint emergency operations (Rutty and Rutty, 2012).

In some respects, maritime emergencies in the Arctic can be considered more demanding than terrestrial emergencies, owing to the complex environment in which they occur. The Arctic Sea region has one of the most sensitive environments on the planet; therefore, any minor incident in this complex environment has the potential to become a major disaster for people, the organizations involved, and the vulnerable marine ecosystem (Strømmen-Bakhtiar and Mathisen, 2012). Long distances, adverse weather conditions, limited emergency response resources (Borch et al., 2016), and heterogenous organizational structures represent just some of the challenges. In a complex environment, inter-organizational collaboration during emergency response tends to become challenging. As a result, there is a need for the clear hierarchical division of tasks, structure, and rapid decision-making processes (Faraj and Xiao, 2006). However, there is an additional need for flexibility, decision-making under time pressure, and informal co-ordination mechanisms (Faraj and Xiao, 2006).

The importance of organizations working with a collaborative perspective while exploring, learning, and building relationships between organizations is highlighted by, among others, Crossan, Lane, and White (1999). In their study, the authors discuss different levels of learning. Moreover, researchers including Crossan, Maurer, and White (2011), Engeström and Kerosuo (2007), Greve (2005), Hardy, Phillips, and Lawrence (2003), Inkpen and Tsang (2007), Jones and Macpherson (2006) and Nooteboom (2008) highlight the need to more studies on inter-organizational learning. Inter-organizational learning processes have become an increasingly relevant field of research, particularly as researchers attempt to understand the context and processes involved in new organizational relationships and settings. However inter-organizational learning related to different settings is poorly investigated (Crossan, Mauer, and White, 2011; Engeström and Kerosuo, 2007; Inkpen and Tsang, 2007; Knight and Pye, 2005; Larsson, Bengtsson, Henriksson, and Sparks, 1998). Therefore, we extend the literature on inter-organizational learning by investigating it in the context of collaborative emergency exercises. We further introduce new processes connected to the inter-organizational learning process while building upon the framework of Crossan et al. (1999) and Jones and Macpherson (2006). Our intentions in this exploratory study are to empirically challenge and validate the "Intuiting, Interpreting, Integration, Institutionalizing, Intertwining" (5I) framework and develop theoretical nuances that enrich our overall understanding of inter-organisational learning processes. For this purpose, we study ECEs in the Arctic. In our model, emergency collaboration exercises are context-sensitive. The ECEs in the Arctic can influence inter-organizational learning depending on the complexity of the external environment.

Although interest in the learning dimension of exercises has grown in recent years (Berlin and Carlström, 2008, 2011, 2014, 2015; Kim, 2013; Perry and Lindell, 2003; Roud and Gausdal, 2019), a general study connecting collaboration exercises and the inter-organizational learning process has remained elusive. In this context, there is a need to develop theoretical and empirical reflections and conduct more in-depth studies in the field of inter-organizational learning. The present study is based on the assumption that inter-organizational learning is understood as part of the continuum of organizational learning proposed by Crossan et al. (1995), Bapuji and Crossan (2004), Holmqvist (2009), Knight (2002), Knight and Pye (2005), and Crossan et al. (2011). Following this line of thought, the present study explores how the inter-organizational learning process can occur from emergency collaboration exercises within a complex environment by building upon Jones and Macpherson (2006). Moreover, we offer a preliminary list of facilitators and impediments of learning processes by studying ECEs in the complex environment of the Arctic.

Theory

Learning

Learning is considered a multi-dimensional phenomenon and can be described as processes that occur at different levels, where learners could be individuals, groups, entire organizations, or inter-organizational networks (Tynjälä, 2008). Learning through emergency exercises is seen as being situated in social contexts, meaning that it occurs through processes of legitimate peripheral participation (Sommer and Njå, 2012).

Learning at the individual level is defined herein as the acquisition of new knowledge (Sommer and Njå, 2012). Two major interpretations of individual learning have been identified by scholars (Becket and Hager, 2002; Harel and Koichu, 2010; Malloch, Cairns, Evans, and O'Connor, 2010). *The individual cognitive approach to learning* focuses on individuals as learners, where learning is understood as the acquisition of information and reasonable behaviour (Baddeley, 1999; Bandura and Walters, 1977; Ormrod, 2008; Piaget, 1972; Skinner, 1965). *The sociocultural approach to learning* focuses on the social relations between people rather than on the individual in isolation (Gherardi, Nicolini, and Odella, 1998). Hence, learning from emergency exercises is considered to be situated in and occurring through processes of participation in various activities and interactions between colleagues (Billett, 2010; Collin, 2002; Eraut, 2007; Lave and Wenger, 1991; Wenger, 2010). Several definitions of group learning were found after reviewing the existing literature. This study will use the definition by London, Polzer, and Omoregie (2005: p. 114), who define group learning as "the extent to which members seek opportunities to develop new skills and knowledge, welcome challenging assignments, are willing to take risks on new ideas, and work on tasks that require considerable skill and knowledge".

Extensive literature reviews have been conducted about organizational learning with multiple conceptualizations (Crossan, Lane, White, and Djurfeldt, 1995; Easterby-Smith, 1997; Huber, 1991; Jones and Macpherson, 2006). The general definition by Huber (1991) is our point of departure toward understanding organizational learning: "an organisation learns if any of its units acquire knowledge that it recognises as potentially useful for the organisation" (p. 126). This definition is valuable because it avoids the assumption that learning inevitably leads to changes in mind and behaviours. However, this definition does not reflect on the process aspect of learning and does not explain when and how obtained knowledge is useful (Crossan et al., 1995; Torres and Preskill, 2001). Therefore, to be more specific, the present study follows the cross-level process approach that assumes that organizational learning is a multi-level process linked through psychological and social processes (Crossan et al., 1999; Bratianu, 2015).

Learning from experiences with other organizations is a major means of organizational learning (Levitt and March, 1988). This experience highlights the importance of organizations working from collaborative perspectives and exploring learning that builds on relationships between organizations (Jones and Macpherson, 2006). This point leads us to the last level – inter-organizational learning – which is a natural result of the growing importance of inter-organizational relationships. In recent years, the focus on studies of organizational learning has been shifting to multi- and inter-organizational learning (Mozzato and Bitencourt, 2014). Inter-organizational learning can be seen as the collective acquisition of knowledge between groups of organizations, thereby compassing the notion of interaction between organizations (Larsson et al., 1998). Therefore, inter-organizational learning is distinct from organizational learning in that it includes the effects of interaction between organizations, which generates synergy and fosters learning (Mozzato and Bitencourt, 2014). Moreover, organizations tend to learn from the experiences of others rather than from their own experience (Perry, 2004). However, inter-organizational learning is supported by organizational processes of knowledge creation and retention (Greve, 2005).

Collaboration is considered important in inter-organizational learning and helps to resolve intractable problems (Jones and Macpherson, 2006). The general aim of collaboration is to provide organizations with a platform for the exchange, transformation, and creation of knowledge. Participating in a collaborative network enables organizations to cross boundaries between different organizations and fields of expertise (Tynjälä, 2008). Moreover, Fayard et al. (2008) believe that it is the collaboration between organizations, which is not limited to organizational boundaries, that gives rise to collective learning.

Multi-level Framework of the Inter-organizational Learning Process

To date, the organizational learning literature had failed to integrate prior research at different levels of analysis (Glynn, 1996; Huber, 1991; Kim, 1998; Nicolini, Crossan, and Easterby-Smith, 2000) until Crossan, Lane, and White (1999) developed a framework that illustrates the processes of learning and how it evolves and is incorporated within organizations. The framework contains a multi-level view of learning and consists of different learning processes that occur within an organization, such as intuiting, interpreting, integrating, and institutionalizing. This study follows the multi-level view of learning because insights and ideas occur in individuals and not organizations (Nonaka and Takeuchi, 1995a; Simon, 1991). Nevertheless, knowledge of the individual does not independently come to bear on the organization. Instead, ideas are shared between individuals, with actions being taken and mutual understanding being developed (Daft and Weick, 1984; Huber, 1991; Schön and Argyris, 1996; Stata, 1989). Complex organizations are more than ad hoc communities or collections of individuals (Crossan et al., 1999). Relationships become structured, and some of the individual learning and shared understandings developed by groups become institutionalized as organization artefacts (Shrivastava, 1983). Crossan et al. (1999) named this multi-level framework the "4I Framework". Within this framework, four processes connect the individual, group, and organizational levels of learning (Crossan et al., 1999). The individual level is based on the learning processes of intuiting and interpreting, while interpreting and integrating are present at the group level. Finally, at the organizational level, integrating and institutionalizing occur.

Crossan et al. (1999) defined *intuiting* as a subconscious process that occurs at the individual level. They argued that this is the beginning of learning and is bound to happen in a single mind. Moreover, intuiting learning involves forming personal experiences. Nonaka and Takeuchi (1995b) stated that this intuition is something that appears before individual actions and is difficult to share with other individuals. *Interpreting* is the second learning process, which Crossan et al. (1999) defined as the conscious elements of individual learning that are shared in groups. *Integrating*, which is the third learning process, is defined as the change of collective understanding at the group level, which functions as a bridge to the organizational level. In this learning process, they argued that the development of shared understanding between individuals occurs and that a change in action is based on mutual adjustments. Crossan et al. (1999) also stated that conversation and joint action are essential for the development of shared understanding. They further elaborate that the integrating process will be informal at the beginning. However, if the change of action repeats itself and is noteworthy, the action will be institutionalized. The last learning process of the 4I model is *institutionalizing*. Crossan et al. (1999) defined institutionalizing as the process where learning is incorporated across the organization. This process works by embedding learning into the organization's systems, structures, routines and practices. The process of institutionalizing is dependent on the defined tasks, specified actions, and organizational mechanisms implemented so that the learning can be put into action (Crossan et al., 1999).

The individual and group learning outcomes that ultimately occur in the body of the organization result in a consensus among members of the organization. Thus, the description of the learning process in integrated organizations is created from individuals, groups, and organizations. In the Crossan, Lane, and White (1999) framework, feedforward learning progresses from individuals' intuiting processes, through group interpretation and integrating, to institutionalizing at the organizational level. Feedforward learning enables the crafting and assimilation of new solutions and is the primary mechanism for organizational adaptation. In feedback processes, learning that has become institutionalized guides (or restricts) future individual and group learning, helping organizations (firms) to exploit their existing knowledge. Notably, both feedforward and feedback mechanisms are required for an organization to benefit from learning (Crossan, Lane, and White, 1999; March, 1991). However, Zietsma et al. (2002) criticise the Crossan et al. (1999) framework by claiming that the exploitation of institutionalized learning is only efficient under stable conditions. However, shared cognitive maps limit the ability of group members to notice and interpret discrepant information (Ansoff, 1977; Bettis and Prahalad, 1995), thereby reducing the organization's adaptability. When the environment changes, reliance on existing knowledge can suppress individual intuiting and/or block it from feeding forward through the group and organization levels of learning.

Zietsma, Winn, Branzei, and Vertinsky (2002) added two new concepts to the original 4I framework. First, *"attending"* captures a more active process of information seeking than the framework for the passive term *"intuiting"* from Crossan et al. (1999), while *"experimenting"* is described as a parallel activity performed by individuals and groups that adds substance to the process of interpreting. Both Zietsma et al. (2002) and Crossan et al. (1999) considered organizational learning processes at three levels of analysis (individual-group-organization) and elaborated on the importance of the external environment to these processes. Later, Jones and Macpherson (2006) extended the 4I framework to what they call 5I framework by including the inter-organizational level and adding *intertwining* as the fifth process (the fifth "I"). The term "intertwining" indicates active engagement between an organization and its external knowledge network. The concept of "intertwining" indicates that learning mechanisms are at the interstices between organizations, and not just within organizational boundaries.

While this framework provides a good understanding of the main processes of the 5I, the present study intends to understand the range and scope of the framework. Part of the study involves scrutinizing the concept to understand the boundaries of the 5I framework. Is the 5I framework complete or can it potentially be extended with the given empirical data? In this study, we explore the inter-organizational learning processes framework in the complex environment of the Arctic by studying collaborative emergency exercises.

In rapidly changing situations within vulnerable and complex environments such as the Arctic, collaboration is not as dependent on a formal structure as it is on ongoing activities that occur in response to future collaboration challenges (Bouty et al., 2012). Owing to a lack of support resource availability and a harsh environment, Arctic emergency management organizations must support each other and develop a collaborative approach towards treating emergencies in the region.

Notably, Hogarth and Makridakis (1981) discuss "competitive" (in this case, "challenging" may be more suitable) and "turbulent" when referring to the complex environments (Hogarth and Makridakis, 1981), with the effects of decisions being *"difficult to predict"*. Hogarth and Makridakis (1981) refer to Slovic, Fischhoff, and Lichtenstein (1977), who suggest that calculating an optimum strategy in a complex environment is challenging. Likewise, in the case of a large-scale emergency in the Arctic, it would be challenging to predict who the participants of a response operation would be, what expertise they have and what further expertise would be required. The Arctic context amplifies challenges related to the aforementioned factors, owing to extreme climate and weather conditions combined with long distances and sparsely populated areas. As a result, Arctic maritime emergency response actions are recognised as particularly challenging jobs that demand well-trained emergency personnel (Borch and Andreassen, 2015).

Emergency Collaboration Exercises in a Complex Environment

Emergency collaboration exercises are unique when it comes to functionality, strategies, and objectives. Objectives can be strategic, focused on practical knowledge building and/or on improving inter-organizational collaboration. ECEs are one of several types of exercises that have been highlighted by academia. S*trategic exercises* aim to simulate an event to examine the results that different interventions can have (Berlin and Carlström, 2015). Thus, the key aim of these exercises is to study the outcomes of different approaches under different conditions and not to increase the learning of tactical level personnel (Babus, Hodges, and Kjonnerod, 1997). *Drill exercises* aim to strengthen individuals' "knowledge in the practice of their profession" (Berlin and Carlström, 2015) and are suitable for tactical- and operational-level personnel to repeat significant elements. *Collaboration exercises* aim to bring different organizations together to integrate actions across organisational boundaries (Berlin and Carlström, 2015) and may be a combination of strategic and drill exercises. From the learning perspective in emergency management, collaboration exercises develop individual, group, and organizational skills by strengthening leadership and triggering inter-organizational curiosity (Andersson et al. 2014). Collaborative interactions between organizations can foster inter-organizational learning, which can occur through a range of inter-organizational activities such as collaborative exercises. Therefore, the emphasis of this study is on the last exercise strategy: "*collaborative exercises*".

Notably, the context of a complex Arctic environment can demand increased collaboration. To better understand the significance of a complex environment in combination with ECEs, a clearer theoretical understanding of the environment and its complexity is required.

We base our definition of environment on Dooley (2004); therefore, we consider the environment as a network of external organizations and institutions (i.e. other agents), as well as the physical surroundings (i.e. resources) (ibid.). The environment can both provide the potential to learn from externals, yet it may also mean a potential to "outsource" and rely on someone else to specialize in specific tasks (Moynihan, 2009). A few aspects of the physical environment become particularly important for emergency exercises. Familiarization with the geography of local surrounds and facilities is important for emergency services, whose core role is an emergency response. However, familiarization with the complexity of a particular environment might not occur automatically (Renner, 2001).

In the present study, complexity characterizes the environment to which organizations and individuals within the organizations are exposed to. We use the definition of complexity proposed by Erdi (2008: p. 7), who defines it as a system where "*circular causality, feedback loops, logical paradoxes and strange loops*" appear. Additionally, the system could be affected by the fact that a "*small change in the cause implies dramatic effects, emergence and unpredictability*".

As a result, based on Dooley (2004) and Erdi (2008), we can define a complex environment as an organization's network of external organizations and institutions (i.e. other agents) as well as the physical surrounding (i.e. resources) that affects the organization, owing to *"circular causality, feed-back loops, logical paradoxes and strange loops"* and the fact that a *"small change* [may imply] *dramatic effects, emergence and unpredictability"*. As optimizing any type of strategy in a complex environment is not an easy task (Hogarth and Makridakis, 1981), this study assumes that the organizations intend to prepare emergency collaboration through inter-organizational learning processes. This could have led to the development of ECEs, in which the personnel of different organizations must interact within complex environments.

Based on the research question and the presented theories, we have developed an analytical model to illustrate the relationships between ECEs and the inter-organizational learning process in a complex environment. Figure 14.1 presents the main elements of this study, where the Arctic context – in the form of an unpredictable and harsh environment with scarce resources and limitations in communication infrastructure – influences collaborative exercises and may affect inter-organizational learning. Additionally, ECEs themselves could influence inter-organizational learning processes.

Methods

Studying learning in real emergency incidents with intensive human interaction is very challenging. We focus on emergency exercises that are more accessible to gather data and study learning processes. In line with a wide range of previous empirical research within emergency management, a case study approach was chosen (Bharosa et al., 2009; Schmied et al. 2017; Sommer and Njå, 2012; Woltjer et al., 2006). ECEs usually produce heterogeneous data in terms of the type of source, the extent of sources and the intended consignee. Moreover, data can sometimes be "classified" or closed to the public. Consequently, a fully embedded case study design is not attainable, owing to incomplete units of analysis in each case (Yin, 2013). Hence, in accordance with Yin (2013), and in contrast to single-case studies in emergency management (Sommer and Njå, 2012), a larger number of similar cases (likely producing similar results) was chosen to overcome this potential

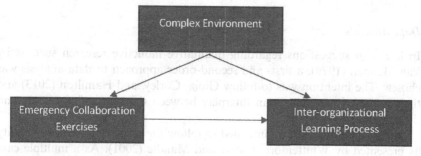

Figure 14.1 Main Elements of the study

weakness with the aim of conducting a more generalizable study (Herriott and Firestone, 1983).

Data Collection

The study focused on four cases of emergency management exercises in the Arctic. Particular focus was given to examine available information on recent full-scale and table-top exercises. The study consists of exercises derived from several large research and development (R&D) projects endorsed by research groups and practitioners in the Arctic (see Table 14.1). The four cases have been selected from a preliminary study of 11 cases.

Choices related to which data to generate and collect were based on how well the ECEs resembled emergency management scenarios in a relevant complex environment. Selection criteria included that the exercises were recent, large scale, multi-organizational, connected to maritime issues, and developing Arctic emergency management competence. A pool of researchers in the area of emergency management used their contacts to search for and gather obtainable data. The remaining four exercise cases are rich in data and focus on collaboration within a complex environment.

The study uses methodological triangulation, with data collection consisting of a set of qualitative methods including archival data from emergency organizations, such as logs and reports, publicly available reports, and presentations. However, the main pillar comprises observation reports, observations, background conversations, and unstructured in-depth interviews related to important exercises. The interviews left room for further questions and detailed inquiries to elaborate on specific elements of the story (Bryman and Bell, 2015).

The study contains data from between 2016 and 2019, when a focus was on maritime exercises located in the Arctic. The data represent mostly complex incident types with a large number of resource requirements and agencies. The exercises included two full-scale exercises – Exercise Nord in 2016 to 2019 and SARex in 2016 – and two simulated table-top exercises: Arctic SAR in 2016 and the Association of Arctic Expedition Cruise Operators (AECO) SAR (Iceland) from 2016 to 2018.

Data Analysis

In line with suggestions regarding qualitative inductive research such as by Van Maanen (1979), a first- and second-order approach to data analysis was chosen. The intention was to follow Gioia, Corley, and Hamilton (2013) and their request for rigour as an interplay between the sources/informants and the researcher.

During data analysis, we intended to follow the primary criteria of validity as presented by Whittemore, Chase, and Mandle (2001). As a multiple case study approach was chosen instead of a single-case study, this helped to

Table 14.1 Description of cases

Name of exercise	Type of exercise	National/ International	Years	Source types	Description
Exercise Nord	Full-scale	National	2016 to 2019	Observation report, background conversation, storybook, interviews	Exercise Nord by Nord University is an annual full-scale exercise that has been taking place for almost 25 years. Every year, the organizers have been able to change the exercise scenario. In 2016, 2018, and 2019 the scenario was connected with an explorer cruise ship dealing with a fire in the engine room and requiring evacuation. In 2017, a terrorism scenario at the university campus was the topic of the exercise. Participants have included Nord University, Salten Fire Brigade, Salten Police District, civil protection, joint rescue, co-ordination centre, Bodø Commune, Nordland Hospital, Coast Guard, Norwegian Society for Sea Rescue RS, Norwegian Air Ambulance, observers, and others.
SARex Exercise	Full-scale	National	2016	Interviews, observation reports, brief and debrief	SARex 2016 was the full-scale exercise in Svalbard connected to testing the implications of the Polar Code on national policies. In addition, practical implications were observed. Participants were University in Stavanger, Norwegian University of Science and Technology, St Olavs Hospital, Nord University, Memorial University of Newfoundland, Arctic University of Norway, Viking Life Saving Equipment, Norsafe, Petroleum Safety Authority Norway, Norwegian Public Authorities, American Bureau of Shipping, and Eni Norge. The goals were to investigate the adequacy of the rescue programme required by the Polar code to study the adequacy of the standard equipment and improve winterization. In addition, the Norwegian Coast Guard personnel were able to share experiences on training on emergency procedures in icy waters, with particular reference to evacuation and rescue from cruise ships.

(continued)

Table 14.1 (continued)

Name of exercise	Type of exercise	National/International	Years	Source types	Description
AECO SAR TTX	Table-top	International	2016 to 2018	Brief and presentation, Observation reports, Interviews	The Association of Arctic Expedition Cruise Operators (AECO) and partners organize annual table-top exercises. Participants have included cruise operators and vessel owners, captains working as masters on different cruise vessels, SAR responders from different entities (in the USA, Canada, Greenland/Denmark, Iceland, Norway, and Finland), the Arctic Coast Guard Forum, observers from universities and research institutions, etc. The exercise has had mostly a workshop character; however, smaller side-workshops have been organized. They were small table-top exercises with simulation elements. Each year, the exercise has been evaluated with a standardized form.
Arctic SAR	Table-top	International	2016	Observation reports	ArcticSAR 2016 was a single-event exercise connected to the seventh work package of the SARINOR project. Participants were mostly from education and research institutions of the SARINOR project, including the Admiral Makarov State University of Maritime and Inland Shipping, Memorial University of Newfoundland, Nord University Business School, and the University of Stavanger. Unique to ArcticSAR was that the exercise participants gathered exercise data from different previous exercises and interview data with professionals and used ArcticSAR as a TTX on a meta-level. This fact meant that observed challenges from previous exercises influenced the scenario of the table-top exercise.

increase the likelihood of decreasing the misinterpretation of "outlier" opinions – hence the increase in *credibility*. Moreover, an in-depth literature review backed the research. Likewise, it was important to juggle the emic perspective of sources – *authenticity* (Whittemore et al., 2001) – related to their own and their organizations' culture with the intended etic perspective of observation and the ultimate aim of this research being applicable – or at least relevant – beyond the culture of the research subjects (Harris, 1976). It helped that the researchers generating the data were from the same project group, which improved the possibility for a comparison of the cases by homogenizing understanding on what to focus on.

The existing literature on 5I learning processes provided the main structure while the data was coded. However, the data itself was the driver for analyzing where the structure from the literature could be extended. Multiple screenings of the coded data and the elimination of non-relevant cases should ensure *criticality*. For example, some of the cases (initially 11, of which four were chosen for the present study) provided data on learning. However, they could either not sufficiently be connected to a complex environment or were not observed by any of the researchers in the closer project team.

Connecting our findings to previous research frameworks and checking back and forth was performed as rigorously as possible as part of the validity control process (*integrity* (Whittemore et al., 2001)). Notably, some aspects of learning (e.g. individual learning) can happen subconsciously and are difficult to articulate (Crossan et al., 1999). Hence, the researchers had to be alert during their observations and then be critical and rigorous to make sense of the coded data.

As previously indicated, the findings were categorized into the processes of intuiting, interpreting, integrating, institutionalizing, and intertwining (Jones and Macpherson, 2006). However, it should be noted that some of the data fit several learning levels and processes. In order to preserve a good overview of the main input from exercises to learning processes, the decision was taken to present the findings in the order of the 5I processes. As a result, the analysis was performed by assessing how the represented data for each processes at each level of learning represented a facilitator or impediment to learning. This is presented in Table 14.2 on "Organisational learning and indicators from ECEs (extended and adapted from Crossan et al. [1999]; Jones and Macpherson [2006]; Dewi, Dwiatmadja, and Suharti [2019])". As a final step, findings giving possible extensions to the 5I framework were stipulated in a separate section (Learning beyond the 5I learning framework). While the first section (Facilitators/impediments) was more empirically driven, the latter (Possible extension of the 5I framework by internalising and interconnecting) emerged while dissecting the concept and the structure of the 5I framework, which was then observed also in a few examples in the empirical data (see sub-chapter "Learning beyond the 5I learning framework").

Findings and Discussion

ECEs and the Arctic

For all the exercises in the present study, the aspect of environmental complexity is omnipresent, if not even the main reason why the exercises are deemed important. All exercises were designed to suit the Arctic environment, and the existence of unpredictability determined the scenarios, owing to a set of unique conditions. These included climatic conditions and social environment, as well as the geographic environment. For example, sources from SARex exercise stated that *"installations can be hundreds of kilometres from shore. At the same time, few or no vessels may be close enough to respond within hours or days"*. AECO (2016) added to this information by stating that *"performing these operations in extreme weather conditions, such as in Polar Regions, presents unique additional challenges, e.g. extremely low temperatures, rapidly changing weather conditions, [the] sparseness of rescue resources, unpredictable presence of sea ice and glacial ice, etc"*.

Particularly in the table top-exercises (TTX), briefs and preparation were used to paint an image by explaining the complexity of the environment. Painting this image was part of the learning process regarding the reality of the context. This included presenting the necessary contextual background information to ensure increased learning effects. Additionally, visual support such as models, pictures, maps, and a movie were used to increase awareness of the complexity of the environment among TTX participants. For the full-scale Exercise Nord, which was designed for students to achieve learning effects, live-streaming and real-time observation information were presented live to students and external observers in order to gauge the multiple challenges which appeared at the same time.

Intuiting and Interpreting

As previously mentioned, intuiting occurs subconsciously and is difficult to observe (Crossan et al., 1999). Discussions with exercise participants before and after Exercise Nord 2019 provided insight into how well the existing collaborative emergency exercises fostered personal competence and skill development. It showed that this large-scale emergency collaboration exercise was mostly accepted as a field to study personal competence at the strategic and political levels. Some of the exercise participants stated that smaller exercises would be able to provide the same or better learning outcomes with fewer resources being required. This could be due to Exercise Nord only happening once a year, during which resources are available than in a realistic scenario for some positions. However, learning on an individual level through an ECE with scenarios that might not be as common as those of smaller exercises can bring benefits to the group, organizational, and inter-organizational levels as well. An example of this was risk management connected to the full-scale ECEs.

Risk management and exercise safety were core topics in the preliminary stages of the exercise and required a good understanding of the challenges of the ECE. Ultimately, participants had to experience a trade-off regarding what would be a completely realistic scenario and ensuring safety. Nonetheless, Exercise Nord 2018 gave indications that exercise participants in education (students) had "good learning" effects. This situation especially relates to practising under time pressure, making difficult decisions, and experiencing a lack of support resources. Also, they actively gathered some tacit knowledge in a largely organised environment specifically designed for them.

During Exercise Nord (2018), we observed the bridge of a distressed vessel and other areas. We observed how the captain and officers had to rely on hand-held telephones for some communication. This was because other information was shared over the loudspeakers and created difficulty in gaining shared information/situational awareness. For individuals, this experience created familiarity with technology and communication tools (Nord, 2018). However, a question remains regarding whether the experienced difficulties from such exercises would create further learning processes on the other learning levels; for example, learning whether another technology (hands-free) was going to be implemented for increased personnel efficiency in vessels of the participating and affiliated organizations. We observed implications for learning processes on communication tools at the group level. The observation revealed that familiarity with technology and communication tools help to establish personal comfort during group interactions. This related to a surprising/stressful experience for individuals. However, there are also reports of overly scripted and easy aspects of the exercises that might be less beneficial for learning at some levels.

On the one hand, the data indicated that the large full-scale exercise type in a complex environment might have some limitations related to intuiting. Several experienced positions from all levels expressed that their role had not contributed to learning on an individual level; for example, "*Captains know [... the] scenario and procedures [rather well before the exercise]*" (Nord, 2017). On other occasions, learning limitations were mostly related to safety reasons and risk management activities. However, challenging activities from a safety perspective could also bring interesting learning effects. Activities such as handling airborne or seaborne resources and transporting casualties must be performed with the utmost care and with knowledge about it from everyone (individuals, groups, organizations, and across organizations). This contribution is noteworthy, as it suggests that in many cases it is not possible to isolate learning effects on one level and instead requires a string of processes connected to learning at all stages.

On the other hand, some interviewees mentioned that the exercise was mostly defined to increase the learning effect for students (in education) and to increase learning effects at the inter-organizational level. This type of statement was mostly connected to when there were roles that were designed to "drill" certain standard operating procedures by the book, which meant limited room for improvization. Some of the more experienced positions actually saw themselves serving support functions in order to make the ECE run smoothly and to reach the exercise goals for students. While this may increase the learning effects for some, others were confronted with inadequate task difficulties. For example, some tasks were too simple for emergency personnel.

Regarding the observed full-scale exercises, exercise support teams were used. These teams were people who would not participate in the actual exercise but would help to facilitate the best learning effects for participants. They had different types of roles to facilitate individual learning processes. While SARex was set up quite strictly before the exercise, the individuals were not guided or steered by the support team during the exercise. During Exercise Nord (2019), this was different for some positions. Some individuals in the exercise had specifically assigned controllers who would be in constant dialogue with each other. The individuals participating in the exercise could come up with questions on what was best, and the controller provided recommendations and feedback to the participants (Nord, 2019). The role of controller facilitated the acquisition of information and the interpretation process, which are the conscious elements of individual learning within group during exercises (Crossan et al., 1999). However, the controllers can additionally be seen as the facilitators of connection between different levels. The controller would bring in knowledge from the group and organizational level, while also being seen as the ambassador feeding back knowledge to the group and organization on how well trained the participants were.

All the exercises in the study provided feedback about how some participating organizations had clear objectives that would influence learning on several levels. Particularly, the two full-scale exercises provided several examples of testing and learning the application of new and innovative equipment. The design of SARex was to test survival equipment. Exercise Nord, however, used a drone, even though feedback showed that further learning on how to operate drones during co-operation was needed. A helicopter pilot stated, *"We are not happy to operate in the same airspace as drones. [We can't] see them because of their size. [However,] clearly, drones have potential and can be used during emergencies"* (Nord, 2019). The pilot was speaking from an organizational perspective, being aware that drones were a risk but had learned and articulated that the benefits should be taken seriously. This type of statement gave further indication of how an individual learning process might also be directly connected with group, organizational, and inter-organizational interests and delivering feedback towards the other levels.

Integrating

Learning on the group level was mostly represented via activities connected to communication and creating mutual understanding during exercises. At

AECO 2016, an interview partner stated that communication personnel were tested, and the whole group level could see how certain staff were best suited for the task at hand. The example was as follows: "*When this guy took over, the information stream was much more clear. That is something that we say to pick your best communication guy on the communications line*". This statement shows how the group learned about certain abilities of one group member. Exercise Nord in 2018 provided further evidence for learning at the group level. It consisted of co-operation between individuals testing their roles. They also learned how to incorporate non-professionals into the exercise. This problem was, in no small degree, at the tactical level. Participants believed that this enabled them to have an idea about the non-professional capabilities of their individual members and organizations. Another example is from the AECO 2018 exercise. The scenario integrated both professionals and non-professional to establish shared views. The following statement shows how the interaction of the participants was intended, although it became clear that some roles had more power to express their opinions than others. One participant said:

A play board [was] used to visualise the vessels in distress during the TTX. The captains were placed around the table with the board. The rest of the group listened to their discussion about what action to take [at] the beginning of the exercise. Afterwards, we were grouped based on what organisation we represented. I played out the role of a passenger. Two persons were running the exercise. It was a good experience for me, although the role of the passengers was not so active; we became more observers of the discussion.

AECO

While room for a constant dialogue with others was slightly limited in the aforementioned set-up (likely owing to people not feeling empowered to participate in the discussions), a good overview of the group's capabilities was created at the individual (interpretation) and group levels (integration). The Data from AECO indicates that group-level learning might always be connected to challenges in providing a framework that enables the active involvement of all individuals in a way that intuition and interpretation are both archived. These examples demonstrate that incorporating members from another background (and from other organizations) also enables organizational learning via evolving an understanding of each other's organizations and creating the necessary trust and predictability.

The discussion with exercise participants at SARex raised further potential challenges to learning on the group level via integration. It seemed that follow-up on exercises and deep conversation on concluding remarks could be challenging. In connection to discussions around Exercise Nord 2019, participants stated that they might return to their own organization's routines after the exercise and did not have the time or resources for the further active integration of learning. They compared their own organization with another organization by stating, *"and I think [the other organisation] [...] is very good with that in comparison with us [...], we are not as good with this"*. By doing so, there is a risk that a learning process could be cut somewhat short at the group- and organizational levels after an ECE, as participants might not further engage with each other in discussions unless follow-up exercises are planned. However, there is often flexibility, in case some participants feel that there is a need for further integration. Data from the SARex exercise revealed that emergency personnel believed that they delivered good results; however, if they thought it was not good enough, they were open to further discussion on how to improve for the next time. This information is based on cognition and only indicates general readiness to follow up.

Although the previous example showed that achieving group learning was difficult, some positions in the Coast Guard gave more insight on potentially successful approaches toward learning. They demonstrated the intention to provide feedback to individuals at the group level. Some people produced reflections regarding what they had experienced before and during the exercise, and then provided recommendations at the group level. One interviewee stated the following:

> The thoughts that I focused on in preparation for the exercise primarily revolved around rationing, distribution of tasks, and watch rotation. Apart from the most obvious challenge related to hypothermia during the exercise, I also became aware of challenges related to socialisation and the importance of including people, maintaining morale, motivation, and communication. Although the most basic physiological needs must be covered in order to survive, I feel that these are also important aspects [...] [on which focus should be placed].
>
> SARex

These thoughts were later shared in reports with all other participants. These reports were then made available publicly and were distributed further within the participating organizations, thereby potentially enabling the institutionalizing process.

Institutionalizing

Institutionalizing includes learning effects within an organization's systems, structures, routines, and practices (Crossan et al., 1999). The Association of Arctic Expedition Cruise Operators (AECO) provided insight into the fact that substantial debriefing and sharing of knowledge (including achievements and lessons learned) could be a method to provide organizational learning effects. However, while we observed substantial effort at the tactical and operational levels, only a few materials were provided regarding high-level organizational discussions on the strategic and political/diplomatic levels

(AECO). Concerning organizational learning about capabilities and how resources might be used, a discussion on where and on which vessels helicopters had the opportunity to land and take off was interesting (Arctic SAR). It seemed that some insights regarding some resources were new to some organizations. One example of this was a discussion about detailed background information regarding vessels as part of the military assets for air patrols. However, while the organizational representatives learned about these resources, it also depends on what the network of the organization could learn from this in the long run and if other individuals and groups from these organizations would be able to access this information if necessary.

The previous example was from a table-top exercise. In terms of debriefing, owing to their tacit character, full-scale exercises can provide an increased potential for learning on tacit experiences compared with table-top exercises. The full-scale Exercise Nord (2019) provided good insights on this topic, as it was a complete exercise in terms of debriefing. It consisted of several debriefing steps at the individual, group, organizational, and inter-organizational levels. Directly after the exercise, professionals had hot wash-ups within their group. (Hot wash-ups are short meetings to discuss the immediate feelings and thoughts after the exercise.) On the same day of the exercise, representatives of all participating organizations met for a joint debriefing and to provide feedback from their organizations to the other organizations. However, the next step in how further learning was distributed afterwards within the institutions could not be assessed. Interviews with participants indicated that this could be more case-to-case oriented and could depend on the available time and resources. However, this also indicates that the total potential for intuiting, interpreting, integrating, and institutionalising would not be available in cases with little time and resources available after exercises.

Data from the SARex exercise suggest that such a commitment to long-term exercise evaluation, learning implementation, and improvement could still be improved. A captain stated the following:

> With the lesson learned here, this phenomenon, as they call it in the Navy is hot wash-up. When they have major naval exercises, no matter how long it takes, they always have such a hot wash-up where they gather all the strengths that have been involved in this exercise/operation, where people give their real opinion on things, and then a report is written afterwards. That is what I say with SAR reporting tools, which I've missed more, that you have to have a way to get a standardised one, where you can go through and take what you did, what you thought was good, what was less good and what you have to learn next time.

The AECO table-top exercises attempted to overcome potential challenges to learning at the organizational level by having exercise participants fill out a form containing questions connected to their perceptions of individual learning, use for organizations, relevance, and recommendations for future exercises. Although this was somewhat subjective – and without sharing too much detail concerning "perceived learning effects" – it ensured that most participants gave feedback and reflection that would benefit organizational and inter-organizational learning if the AECO table-top exercise network provided in-depth insights with other inter-organizational networks.

In contrast to the periodic recurrent exercises, such as those for AECO and Nord, Arctic SAR TTX was a single event. The exercise produced feedback for both maritime sector-related R&D reports and recommendations for policy and academia. Nonetheless, challenges were still connected to the participating group, such as being isolated from the actual organizations that they were discussing. As a result, the exercise could not guarantee learning at the organizational level. Furthermore, it remained unclear whether the aforementioned reports had sufficient power to potentially change the organizational structure, routines, and procedures according to the exercise outcomes and evaluation to produce long-term learning effects.

Intertwining

Learning on the inter-organizational level was connected to intertwining, which is an active engagement between the organization and its external knowledge network (Jones and Macpherson, 2006). Notably, AECO 2016 provided a great example for active engagement across organizations. They established a resilient inter-organizational trust aspect in their exercise goals, stating that "*the objective for this workshop and TTX is to strengthen the cooperation and exchange of knowledge between the Arctic cruise industry and various Arctic SAR responders*" (AECO Reykjavik 2016). Already, the sheer participation of a broad group of organizations at the AECO exercise could be seen as an indicator for increased intertwining. However, the data from AECO indicated that exercise participants from a meta-organizational level (the Arctic Coast Guard Forum in this case) demanded further "*sharing information and best practice [as well as to] encourage more exercises and the systematic sharing of lessons learned*" (AECO 2018). Hence, this seems to be an indication that inter-organizational learning depends on the increased professional collaboration of competent exercise participants who can then contribute to the inter-organizational level of learning.

If these competent people are not participating, this was a factor that was raised as an impediment to learning. Certain stakeholders who were deemed necessary to create further inter-organizational learning did not participate in the exercise. For example, one exercise participant highlighted that "*there were few participants from the industry*" (AECO 2016).

The international TTX set-up of AECO 2018 demonstrated the importance of participation and exchange by a wide variety of actors to provide learning on the inter-organizational level. Participants were eager to learn about tasks and restrictions connected to co-operation with different organizations. The participation of different coast guards, different institutions, and different nationalities represented an example of what the actual challenges and hindrances of a real case could look like. For example, everyone wanted to know *"What role does* [the Joint Rescue Coordination Centre of country A] [...] *play at this stage?"* in a scenario presented by AECO.

Another factor that can be connected to "intertwining" is how far learning at the inter-organizational level can be spread beyond the organizations participating in the exercise.

For example, during the table-top exercises and particularly at the META TTX Arctic SAR, the setting gave room for participants to discuss and be exposed to different views and approaches. This situation was due to the design of the exercise, where experience from observations of previous exercises, findings from interviews, previous work packages, and analyzed incidents were used to investigate gaps in training, education, and collaboration across institutional borders (Arctic SAR). However, while this produced the potential for learning at the inter-organizational level connected to the "intertwining" of organizations, the learning effect at the individual and group levels remained limited. How much the learning effect would spread among the discussed organizations was not assessable, as it was a once-off event.

In contrast, an aspect that seemed to provide an additional inter-organizational learning effect was the periodic recurrence of Exercise Nord. Each year, stakeholders participate in a discussion to organize the next exercise based on the learning gaps from the previous year, as well as on what their organizations wanted to be trained on. Similarly, the AECO exercises also produced recurrent feedback, owing to similar stakeholders gathering every year. In addition, press releases contained lessons learned and recommendations, such as AECO's report that made learning outcomes available beyond the participating organizations; however, what effect these materials have had cannot be measured.

Analysis of the 5I Framework

The analysis was performed by assessing how the data for each 5I process represented a facilitator or impediment to learning. In the process of data analysis, the secondary data and coded qualitative data from the observations and interviews were merged and analyzed as a whole (Mays and Pope, 2000; Miles, Huberman, Huberman, and Huberman, 1994). Similar to the approach taken by Dewi et al. (2019), Table 14.2 represents an adaptation to Crossan et al. (1999). However, it is extended by the inter-organizational level and intertwining process related to the 5I learning framework (Jones and Macpherson, 2006). Also, through the data, we were able to determine the value for "input/output" at the inter-organizational level.

In contrast to Dewi et al. (2019), who established *"determination"*, Table 14.2 is extended by two columns representing the codes related to indicators from the data. In line with the terminology in other case studies such as Zietsma et al.

Table 14.2 Organizational learning and indicators from ECEs (extended and adapted from Crossan et al., 1999; Jones and Macpherson, 2006; Dewi et al., 2019)

Learning level	5I process	Input/outcomes	Learning facilitators	Learning impediments
Individual	Intuiting	Learning through experience, images, metaphors	Opportunities to make mistakes Ability to test different strategies in exercise The possibility to be exposed by an alternative view Personal competence and skills development Practice taking action under pressure during exercise Familiarity with technology and communication tools that helps to establish personal comfort Exercises allow you to get your hands dirty	Acting in isolation and passively Individuals tend to focus on their own issues The task difficulties are not adequate. Some tasks are very simple Predefined roles and tasks with limited improvization Challenges in transferring the experience to colleagues Lack of structured self-evaluation
	Interpreting	Learning through the cognitive map, dialogue, reflections	Openness to divergent view Testing innovative approaches Not shy to ask for guidance Constant dialogue among individuals Practice professional language of emergency response	

Learning level	5I process	Input/outcomes	Learning facilitators	Learning impediments
Group	Integration	Group confirms or changes individual learning Learning through shared understanding, mutual adjustment, interactive system, and reflections	Discussion after exercises Realistic scenario makes exercise participant learn the most Application of systematic approach and guideline facilitates learning Joint-sensemaking during exercises Learn how to co-operate and follow the command in the exercises Exercises establish shared view in temporary groups Interactive dialogue within the group	Following up the exercises and a deep conversation on concluding remarks is a challenge, as participants get back to their own organization's routines after the exercise
Organization	Institutionalizing	Learning through organisation change or confirmation of group learning and making decisions on redefining existing procedures, routines, rules, and structures	Very few high-level organizational discussions Each organization has its own hot wash-up, so personnel who did not involve in joint preparation phase activities will hear about other organizations' competence	Isolation of the group that attended exercises within the organization Lack of commitment to change in organizational structure routines and procedures according to exercise outcomes and evaluation Poor incorporation of debriefing and low priority of evaluative learning Resistance towards changing organizational culture

(continued)

Table 14.2 (continued)

Learning level	5I process	Input/outcomes	Learning facilitators	Learning impediments
Inter-organization	Intertwining	Learning through inter-organizational response facilitation, networking, and inter-organizational trust	Developing competences through inter-organizational collaboration, especially regarding international rules and regulations Developing relationship across institutions and organizational borders Organizations establish resilient inter-organizational trust by involvement in exercises	Different organizational culture or restrictions in the military, civilian, and volunteer organizations Lack of continual evaluation and reassessment of developed relationships

(2002), we called those indicators facilitators and impediments. Each of the facilitators and impediments in the table have a code corresponding to what is described in the findings and analysis in the 5I process. The analysis of 5I the framework in this study confirmed that the processes are possible to recognize at an inter-organizational level. However, we observed the potential to expand the framework by adding two more processes at the group and inter-organizational level based on the empirical data (this is explained in the next section on "Learning Beyond the 5I Learning Framework").

Learning Beyond the 5I Learning Framework

The approach of this study was to explore the processes of the 5I learning framework in the context of ECEs in a complex environment. However, the study was able to identify learning effects that could go beyond the 5I framework. Table 14.3 illustrates the elements that the 5I framework has covered across learning levels and to what extent they have been covered. Several of the connections beyond the 5Is (suggested with dashed lines in the 5I model) are possible. As the main example, the effects of periodical recurrence (introduced in the findings on intertwining) indicate that the other levels (individual, group, and organization) could also benefit from the periodic recurrence of the exercises.

The yellow boxes in Table 14.3 reveal that the 5I framework covers learning among the group levels only to a minor degree (Jones and Macpherson, 2006). At this level, we recognized the potential to expand the framework by adding a process.

We observed that groups from the same organizations from different levels and departments learned how to co-operate and communicate. This learning was real, based on response groups from different organizations as well. For example, the on-scene personnel from the fire brigades closely interacted with the Coast Guard personnel during the Nord Exercises (Nord 2016, 2018, 2019). Notably, a form of inter-group collaboration enabled participants to learn efficient ways of working together by establishing mutual understanding over a short time period for emerging temporary organizations (including groups from different organizations or the same organizations). The data in this study support the learning occurring between groups to some degree; however, further quantitative data will be required to fully support this idea. We called this an *internalizing* process because the group established a swift understanding and transferred information internally between themselves during the emergency response in the context of the exercises.

The yellow boxes in Table 14.3 also reveal that the 5I framework covers learning among the inter-organizational levels to only a minor degree. At this level, we recognized the potential to expand the framework by adding a process. Our literature review on previous studies had suggested that the inter-organizational level was only a sub-group of the organizational level, to some

Table 14.3 Potential for expanding the 5I framework

Learning level	Individual	Group	Organization	Inter-organization
Individual	Intuiting (5I)	Interpreting (5I)	Suggested with dashed lines within the 5I learning framework but not discussed in detail	Suggested with dashed lines within the 5I learning framework but not discussed in detail
Group	Suggested with dashed lines within the 5I learning framework but not discussed in detail	Not suggested as one of the 5I	Integrating (5I)	Suggested with dashed lines within the 5I learning framework but not discussed in detail
Organization	Suggested with dashed lines within the 5I learning framework but not discussed in detail	Suggested with dashed lines within the 5I learning framework but not discussed in detail	Institutionalizing (5I)	Intertwining (5I)
Inter-organization	Suggested with dashed lines within the 5I learning framework but not discussed in detail	Suggested with dashed lines within the 5I learning framework but not discussed in detail	Suggested with dashed lines within the 5I learning framework but not discussed in detail	Not suggested as one of the 5I

extent. However, within the context of environmental complexity, it was shown that there is potential for inter-organizational-level learning to be fostered through ECE. Some research from the area of a sociocultural approach to learning could provide a conceptual background to this learning level. However, this is missing empirical support (Mozzato and Bitencourt, 2014). The data revealed that in some exercises (such as those by AECO), participants were from different networks, meaning that different emergency management networks gathered together to learn from each other. This learning facilitates communication and familiarizes them with other structures and working procedures. In other words, they learn from being connected to a larger network. We called this the *interconnecting* process – the learning process that occurs between inter-organizational networks. Figure 14.2 presents our extension to the framework with the addition of internalizing and interconnecting processes.

Conclusion

In this study, we used the 5I framework to analyze the learning process in the context of collaborative exercises. We assessed the suitability of the 5I framework for understanding inter-organizational learning processes in emergency management in general and collaboration exercises in particular. The 5I

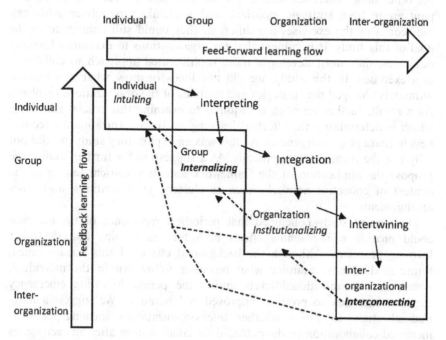

Figure 14.2 The 7I learning framework adapted from Jones and Macpherson (2006).

framework was initially developed in the context of small and medium-sized enterprises that operate in mature manufacturing sectors. Notably, the application of this framework to emergency collaborative exercises can be considered as an innovative approach to understand learning at the inter-organizational level more generally. We contributed to the framework by proposing two new processes: *internalizing* and *interconnecting*. *Internalizing* occurs between groups of the same organization or different organizations, while *interconnecting* occurs at the inter-organizational level. Apart from extending the framework, our study identified preliminary learning facilitators and impediments in the complex environment of the Arctic during ECEs. Overall, this study shed light on how the complex environment of the Arctic influences ECEs, as well as learning processes.

Moreover, the study illuminated how ECEs can affect inter-organizational learning processes to some degree. Notably, this study has several practical implications. For example, the results can be useful for exercise organizers, those who invest in exercises, and the emergency personnel who participate in exercises. Owing to the qualitative research design and the heterogeneity of the data, the generalization of these results must be done carefully.

On the one hand, the full-scale Exercise Nord, with over 1,000 participants, had a vast participant field ranging from tactical and operational to strategic and political levels, which went through the learning processes. On the other hand, exercises such as Arctic SAR had a much smaller participant group from strategic, political, and academic perspectives. Although the scopes of the exercises are different, they could still contribute to the goal of this study. It is evident that for organizations to maximize learning outcomes, they must develop a more sophisticated approach to collaboration exercises. In this study, we did not focus on how what was learned ultimately changed the strategies and routines of the organizations involved. As a result, further research is required to examine this crucial next step, which is determining the effect of learning on the organizational effectiveness in managing emergencies. As this was an exploratory study, we did not fully test the framework empirically. As a suggestion for further studies, we propose the application of the framework and its empirical testing in the context of collective networks such as clusters, joint ventures, and other arrangements.

The present study concluded that periodical recurrence of the exercises could provide wide-reaching effects both for intertwining and the other learning processes. Although the background effects of this remain veiled, future studies must examine what processes follow within the individual, group, and organizational levels during the period following emergency response exercises, to promote improved collaboration. We suggest a longitudinal study to assess whether inter-organizational learning leads to increased collaboration or the potential for collaboration after an exercise, as well as what processes are enabled by periodic response exercises.

References

Andersson, A., Carlstrom, E.D.Ahgren, B., and Berlin, J.M. (2014). Managing boundaries at the accident scene–a qualitative study of collaboration exercises. *International Journal of Emergency Services*, 3(1): 77–94.

Babus, S., Hodges, K., and Kjonnerod, E. (1997). Simulations and institutional change: Training US government professionals for improved management of complex emergencies abroad. *Journal of Contingencies and Crisis Management*, 5(4): 231–240.

Baddeley, A. (1999). *Cognitive psychology: A modular course. Essentials of human memory.* Hove: Psychology Press/Taylor & Francis.

Bandura, A., and Walters, R.H. (1977). *Social Learning Theory* (Vol. 1). Englewood Cliffs, NJ: Prentice-Hall.

Bapuji, H. and Crossan, M. (2004). From questions to answers: reviewing organizational learning research. *Management Learning*, 35(4): 397–417.

Becket, D. and Hager, P. (2002). *Life, work and learning. Practice in postmodernity.* New York: Routledge.

Berlin, J.M. and Carlström, E.D. (2011). Why is collaboration minimised at the accident scene? A critical study of a hidden phenomenon. *Disaster Prevention and Management: An International Journal*, 20(2): 159–171.

Berlin, J.M. and Carlström, E.D. (2015). Collaboration exercises: What do they contribute? *Journal of Contingencies and Crisis Management*, 23(1): 11–23.

Billett, S. (2010). *Learning through Practice.* Amsterdam: Springer.

Borch, O.J. and Andreassen, N. (2015). Joint-Task Force Management in Cross-Border Emergency Response. Managerial Roles and Structuring Mechanisms in High Complexity-High Volatility Environments. In A. Weintrit and T. Neumann (eds). *Information, Communication and Environment: Marine Navigation and Safety of Sea Transportation:* 217. London: CRC Press.

Borch, O.J., Andreassen, N., Marchenko, N., Ingimundarson, V., Gunnarsdóttir, H., Jakobsen, U., Markov, S. et al. (2016). Maritime activity and risk patterns in the High North: MARPART Project Report 2.

Bratianu, C. (2015). Organizational Learning and the Learning Organization. In *Organizational Knowledge Dynamics: Managing Knowledge Creation, Acquisition, Sharing, and Transformation.* Hershey, PA: Information Science Reference: 286–312.

Bryman, A. and Bell, E. (2015). *Business Research Methods*: New York: Oxford University Press,.

Chen, R., Sharman, R., Rao, H.R., and Upadhyaya, S.J. (2008). Co-ordination in emergency response management. *Communications of the ACM*, 51(5): 66–73.

Collin, K. (2002). Development engineers' conceptions of learning at work. *Studies in continuing education*, 24(2): 133–152.

Crossan, M.M., Lane, H.W., White, R.E., and Djurfeldt, L. (1995). Organizational learning: Dimensions for a theory. *The International Journal of Organizational Analysis*, 3(4): 337–360.

Crossan, M.M., Lane, H.W., and White, R.E. (1999). An organizational learning framework: From intuition to institution. *Academy of Management Review*, 24(3): 522–537.

Crossan, M.M., Maurer, C.C., and White, R.E. (2011). Reflections on the 2009 AMR decade award: do we have a theory of organizational learning? *Academy of Management Review*, 36(3): 446–460.

Daft, R.L. and Weick, K.E. (1984). Toward a model of organizations as interpretation systems. *Academy of Management Review*, 9(2): 284–295.

Dewi, Y.E.P., Dwiatmadja, C., and Suharti, L. (2019). A qualitative study on learning organization as an essential action lowering skill mismatch effects. *Business: Theory and Practice*, 20: 50–60.

Dooley, K.J. 2004. Complexity Science Models of Organizational Change and Innovation. In Poole, M.S. and Van de Ven, A. H. (eds), *Handbook of Organizational Change and Innovation* 374–397. Oxford: Oxford University Press.

Easterby-Smith, M. (1997). Disciplines of organizational learning: contributions and critiques. *Human Relations*, 50(9): 1085–1113.

Engeström, Y. and Kerosuo, H. (2007). From workplace learning to inter-organizational learning and back: the contribution of activity theory. *Journal of Workplace Learning*, 19(6): 336–342.

Eraut, M. (2007). Learning from other people in the workplace. *Oxford Review of Education*, 33(4): 403–422.

Erdi, P. 2008. *Complexity Explained*: 237–303. Berlin/Heidelberg: Springer.

Faraj, S. and Xiao, Y. (2006). Coordination in fast-response organizations. *Management Science*, 52(8): 1155–1169.

Fayard, P. Apresentação. (2008). InA. Balestrin and J. Verschoore, *Redes de cooperação empresarial: estratégias de gestão na nova economia*. Porto Alegre: Bookman.

Gherardi, S., Nicolini, D., and Odella, F. (1998). Toward a social understanding of how people learn in organizations: The notion of situated curriculum. *Management Learning*, 29(3): 273–297.

Gioia, D.A., Corley, K.G., and Hamilton, A.L. (2013). Seeking qualitative rigor in inductive research: Notes on the Gioia methodology. *Organizational Research Methods*, 16(1): 15–31.

Glynn, M.A. (1996). Innovative genius: A framework for relating individual and organizational intelligences to innovation. *Academy of Management Review*, 21(4): 1081–1111.

Greve, H.R. (2005). Interorganizational learning and heterogeneous social structure. *Organization Studies*, 26(7): 1025–1047.

Hardy, C., Phillips, N., and Lawrence, T.B. (2003). Resources, knowledge and influence: The organizational effects of inter-organizational collaboration. *Journal of Management Studies*, 40(2): 321–347.

Harel, G. and Koichu, B. (2010). An operational definition of learning. *The Journal of Mathematical Behavior*, 29(3): 115–124.

Harris, M. (1976). History and significance of the emic/etic distinction. *Annual Review of Anthropology*, 5(1): 329–350.

Herriott, R.E. and Firestone, W.A. (1983). Multisite qualitative policy research: Optimizing description and generalizability. *Educational Researcher*, 12(2): 14–19.

Hogarth, R.M. and Makridakis, S. (1981). The value of decision making in a complex environment: An experimental approach. *Management Science*, 27(1): 93–107.

Holmqvist, M. (2009). Complicating the organization: a new prescription for the learning organization? *Management Learning*, 40(3): 275–287.

Huber, G.P. (1991). Organizational learning: The contributing processes and the literatures. *Organization Science*, 2(1): 88–115.

Inkpen, A.C. and Tsang, E.W. (2007). 10 Learning and Strategic Alliances. *The Academy of Management Annals*, 1(1): 479–511.

Bouty I., Godé C., Drucker-Godard C., Lièvre P., Nizet J., and Pichault F. (2012). Coordination practices in extreme situations. *European Management Journal*, 30(6): 475–489.

Jones, O. and Macpherson, A. (2006). Inter-organizational learning and strategic renewal in SMEs: extending the 4I framework. *Long Range Planning*, 39(2): 155–175.

Kapucu, N., Arslan, T., and Demiroz, F. (2010). Collaborative emergency management and national emergency management network. *Disaster Prevention and Management: An International Journal*, 19(4): 452–468.

Kim, D.H. (1998). The link between individual and organizational learning. *The Strategic Management of Intellectual Capital*, no. 41, 62

Knight, L. (2002). Network learning: Exploring learning by interorganizational networks. *Human Relations*, 55(4): 427–454.

Knight, L. and Pye, A. (2005). Network learning: An empirically derived model of learning by groups of organizations. *Human Relations*, 58(3): 369–392.

Kristiansen, E., Løwe Sørensen, J., Carlström, E., and Inge Magnussen, L. (2017). Time to rethink Norwegian maritime collaboration exercises. *International Journal of Emergency Services*, 6(1): 14–28.

Larsson, R., Bengtsson, L., Henriksson, K., and Sparks, J. (1998). The inter-organizational learning dilemma: Collective knowledge development in strategic alliances. *Organization Science*, 9(3): 285–305.

Lave, J. and Wenger, E. (1991). *Situated Learning: Legitimate Peripheral Participation*. Cambridge: Cambridge University Press.

Levitt, B. and March, J.G. (1988). Organizational learning. *Annual Review of Sociology*, 14(1): 319–338.

London, M., Polzer, J.T., and Omoregie, H. (2005). Group learning: A multi-level model integrating interpersonal congruence, transactive memory and feedback processes. *Human Resource Development Review*, 4(2): 114–136.

Malloch, M., Cairns, L., Evans, K., and O'Connor, B.N. (2010). *The SAGE Handbook of Workplace Learning*. Thousand Oaks, CA: Sage Publications.

March, J.G. (1991). Exploration and exploitation in organizational learning. *Organization Science*, 2(1): 71–87.

Mays, N. and Pope, C. (2000). Assessing quality in qualitative research. *British Medical Journal*, 320(7226): 50–52.

Miles, M.B., Huberman, A.M., Huberman, M.A., and Huberman, M. (1994). *Qualitative Data Analysis: An Expanded Sourcebook*. Thousand Oaks, CA: Sage Publications.

Moynihan, D.P. (2009). From intercrisis to intracrisis learning. *Journal of Contingencies and Crisis Management*, 17(3): 189–198.

Mozzato, A.R. and Bitencourt, C.C. (2014). Understanding interorganizational learning based on social spaces and learning episodes. *BAR-Brazilian Administration Review*, 11(3): 284–301.

Nicolini, D., Crossan, M., and Easterby-Smith, M. (2000). Organizational learning: debates past, present and future. *Journal of Management Studies*, 37(6): 783–796.

Nonaka, I. and Takeuchi, H. (1995a). *The Knowledge-creating Company: How Japanese Companies Create the Dynamics of Innovation*. Oxford: Oxford University Press.

Nonaka, I. T. and Takeuchi, H. (1995b). *The Knowledge-creating Company*. Oxford: Oxford University Press.

Nooteboom, B. (2008). Learning and Innovation in Inter-organizational Relationships. *In The Oxford Handbook of Inter-organizational Relations.* Oxford: Oxford University Press.

Ormrod, J. (2008). Metacognition, Self-regulated learning, and Study strategies. In *Human Learning* (5th ed.): 350–390. Upper Saddle River, NJ: Pearson/Merrill Prentice-Hall.

Perry, R. (2004). Disaster exercises' outcomes for professional emergency personnel and citizen volunteers. *Journal of Contingencies and Crisis Management*, 12(2): 64–75.

Piaget, J. (1972). *The Principles of Genetic Epistemology* (translated by Wolfe Mays). London: Routledge/Kegan Paul.

Renner, S. (2001). Emergency exercise and training techniques. *Australian Journal of Emergency Management*, 16(2): 26.

Roud, E.K.P., Borch, O.J., Jakobsen, U., and Marchenko, N. (2016). Maritime Emergency Management Capabilities in the Arctic. Paper presented at the Paper presented at the 26th International Ocean and Polar Engineering Conference.

Roud, E. and Gausdal, A.H. (2019). Trust and emergency management: Experiences from the Arctic Sea region, *Journal of Trust Research.* doi:10.1080/21515581.2019.1649153.

Rutty, G.N. and Rutty, J.E. (2012). Did the participants of the mass fatality exercise Operation Torch learn anything? *Forensic Science, Medicine, and pathology*, 8(2): 88–93.

Schmied, J., Borch, O.J., Roud, E.K.P., Berg, T.E., Fjørtoft, K., Selvik, Ø., and Parsons, J.R. (2017). Maritime operations and emergency preparedness in the Arctic–competence standards for search and rescue operations contingencies in Polar Waters. In *The Interconnected Arctic—UArctic Congress 2016*: 245–255. Cham: Springer.

Schön, D. and Argyris, C. (1996). *Organizational Learning II: Theory, Method and Practice.* Reading, MA: Addison Wesley.

Shrivastava, P. (1983). A typology of organizational learning systems. *Journal of Management Studies*, 20(1): 7–28.

Simon, H.A. (1991). Bounded rationality and organizational learning. *Organization Science*, 2(1) 125–134.

Skinner, B.F. (1965). *Science and Human Behavior.* : Simon and Schuster.

Slovic, P., Fischhoff, B., and Lichtenstein, S. (1977). Behavioral decision theory. *Annual Review of Psychology*, 28(1): 1–39.

Sommer, M. and Njå, O. (2012). Dominant Learning Processes in Emergency Response Organizations: A Case Study of a Joint Rescue Coordination C entre. *Journal of Contingencies and Crisis Management*, 20(4): 219–230.

Stata, R. (1989). Organizational Learning—The Key to Management Innovation . *MIT Sloan Management Review*, 30(3).

Strømmen-Bakhtiar, A. and Mathisen, E. (2012). Sense-giving systems for crises in extreme environments. Proceedings of InSITE 2012: 91–112.

Torres, R.T. and Preskill, H. (2001). Evaluation and organizational learning: Past, present, and future. *American Journal of Evaluation*, 22(3): 387–395.

Tynjälä, P.J.E. (2008). Perspectives into learning at the workplace. *Educational Research Review*, 3(2): 130–154.

Van Maanen, J. (1979). The fact of fiction in organizational ethnography. *Administrative Science Quarterly*, 24(4): 539–550.

Wenger, E. (2010). *Communities of Practice and Social Learning Systems: the Career of a Concept Social Learning Systems and Communities of Practice*: 179–198. Cham: Springer.

Whittemore, R., Chase, S.K., and Mandle, C.L. (2001). Validity in qualitative research. *Qualitative Health Research*, 11(4): 522–537.

Woltjer, R., Lindgren, I., and Smith, K. (2006). A case study of information and communication technology in emergency management training. *International Journal of Emergency Management*, 3(4): 332–347.

Yin, R.K. (2013). *Case Study Research: Design and Methods*: Thousand Oaks, CA: Sage Publications.

Zietsma, C., Winn, M., Branzei, O., and Vertinsky, I. (2002). The war of the woods: Facilitators and impediments of organizational learning processes. *British Journal of Management*, 13(2): 61–74.

15 Practical Co-operation on Oil Spill Preparedness and Response in the Barents Sea and the Arctic

Ole Kristian Bjerkemo

Introduction

Norway and Russia have a common border in the Barents Sea. This area is very rich with in stocks of fish. This joint stock is vulnerable, e.g. oil spills and a large oil spill in this vulnerable area could lead to significant consequences.

The agreement between Norway and the Russian Federation concerning Co-operation on the Combatment of Oil Pollution in the Barents Sea was signed on 28 April 1994 by ministers from the two states. In addition to the agreement, the Director of Marine Pollution Control and Salvage Administration of the Russian Federation and the Director of the Norwegian Pollution Control Authority signed the Joint Plan on the Combatment of Oil Pollution in the Barents Sea.

In addition to the common border in the Barents Sea, the two states are also Arctic states. Because of that, both Russia and Norway in 2013 signed the 'Agreement on Co-operation on Marine Oil Pollution Preparedness and Response in the Arctic' also named MOSPA.

Implementation of the Agreements

An agreement is not worth much without a thorough implementation. To implement the agreements, the parties to the Barents Sea agreement and MOSPA have regarded regular exercises as very important. Such exercises could be alarm exercises, table-top exercises, operational exercises, or a combination of these.

An important element in case of large oil spills is how the parties should co-operate at sea and between the incident command of the lead country and those countries assisting. The co-ordination at sea is generally through an On-Scene Commander (OSC) from the lead country. The OSC will report to the national incident command in their homes state. As the command system at sea is more or less the same in all states, even a bilateral agreement and/or the MOSPA is activated, it is deemed that operations at sea can be handled relatively unproblematically, regarding command and control. Such operations will definitely be challenging because of other factors such as long distances, darkness in long parts of the year, and cold and ice, in addition to many other factors. The Agreement states:

> Each Party to the Agreement has in place existing command and control systems that are used during oil pollution incidents within the areas under its jurisdiction. There also exist other bilateral and multilateral agreements between Arctic States that establish methodologies for a joint response, in which command and control systems have been predefined. Therefore, it is not advisable to create a common general command and control system for the Parties to the Agreement.

The parties to the Barents agreement have agreed to use the international POLREP system to alarm each other and request assistance. The initial contact points are the Vardø Vessel Traffic Services (Vardø VTS) in Norway and the Marine Rescue Co-ordination Services in Murmansk. The system for notification and request for assistance related to the MOSPA is different. There are two systems for notification in place, and they have similarities, but two systems could lead to challenges.

Exercises

For both the bilateral agreement between Norway and Russia and the MOSPA, the starting point for an exercise is a planning conference. Close contact during the planning is important for understanding and learning from each other. The experience from the planning of exercises is that the planning itself can give more learning experiences than the exercise itself. Through the planning process, the involved parties will normally develop exercise objectives and other important elements.

Because of increased oil exploration in the Norwegian sector of the Barents Sea, in 2016 Norway initiated table-top exercises to clarify co-operation between the parties in the event of massive oil spills from the oil and gas activity. One important reason for this was that a spill from this activity would drift into the Russia exclusive economic zone within a few hours. Through these exercises a need was identified for improvement in the Joint Plan between Norway and Russia. It was also identified that the system for command and control of the response for the private sector might be different from how the governmental institutions have organized this. Another important result of these exercises was the establishment of a Joint Co-ordination Centre (JCC). This centre should be active if an oil spill were to affect both Russian and Norwegian territorial waters and shorelines. In the Joint Plan, the following text describes the actual situation:

> In a situation where oil is drifting at sea and has beached on both sides of the border, the parties may establish a Joint Co-ordination Centre (JCC). The JCC might be established in a central place of the affected area.

The main objectives of such a centre are to ensure good coordination and information exchange between the parties.

In a report from the third oil spill response exercise between Norway and Russia in 1997, it was recorded that:

> The exercise proved satisfactory and confirmed that the Norwegian and Russian oil spill organizations could operate together. The exercise showed the necessity for more joint exercises in order to clarify and disclose any weaknesses in the oil spill response plan to attain maximum utilization of response resources in the Barents Sea.

The results were very positive and promising for future co-operation. The parties were able to co-operate because of a well-known command and control system for the response at sea.

From the annual joint exercise in 2018, we can see that there are still lessons to be learned. Nevertheless, we can also see that important elements for operational co-operation works well, e.g. the command and control system. One of the bullet points from the exercise is:

- Effective on-scene co-ordination and use of resources in Oil Spill Response operations.

Conclusion

2019 marked the 25th anniversary of the signing of the oil spill agreement between Norway and Russia on the Barents Sea. That agreement and the MOSPA agreement have not yet been tested in a real spill situation, only through exercises. The results from the exercises are very promising, and it seems that the common command and control system for response at sea works well. It is anticipated that a joint shoreline response operation would be much more complicated.

Through the exercises, the parties identified improvements, which have been implemented in joint contingency plans and operational guidelines. These improvements are related to tactical, operational, and strategic issues.

16 Search and Rescue in the Arctic

Petr Gerasun

According to the theoretical conclusions of scientists and researchers, global warming that causes significant levels of Arctic ice to melt is not that far off. However, we have to provide search and rescue (SAR) of people in distress at sea in the here and now, including in challenging icy waters. If we look at the picture of ice formations in the Arctic, we can observe that ice of 10–30 cm in thickness was formed in early November 2019 along almost the whole Northern Sea Route (NSR)[1].

What does it mean for us? It means that all shipping along the NSR is possible with the use of icebreakers. All Maritime Rescue Co-ordination Centers (MRCC) and Maritime Rescue Subcenters (MRSCs) of the Russian Federation implement the so-called 'Basin's plans'. These plans specify actions of organizations employing SAR resources in missions to rescue people in distress at sea and to respond to oil spills. Consequently, the ice-breakers, which assist vessels along the NSR, perform SAR functions at the same time. One or several medical professionals, rescuers, oil spill response (OSR) equipment and specialists, sometimes helicopters, are on board the icebreakers and ready to assist in missions. Some stores with the necessary OSR equipment are located onshore along the NSR. During the ice-free season, there is quite a busy shipping activity along the NSR, and maritime rescue multi-purpose tugs with the necessary SAR and OSR equipment and facilities on board keep watch in located points.

What are the challenging issues? There are a small number of healthcare facilities in the Arctic, and hospitals are mostly situated in major population centres. Therefore, the transportation of a sick or injured person to the hospital can take a long time. The transfer of the necessary equipment to an emergency site is often possible only using a helicopter. which might be located far away. As a consequence, a large area and long distances add to the time to evacuate people and transfer the necessary equipment compared with mainland missions. However, during the ice season, the number of vessels falls, and as a result, the number of emergencies decreases in this period.

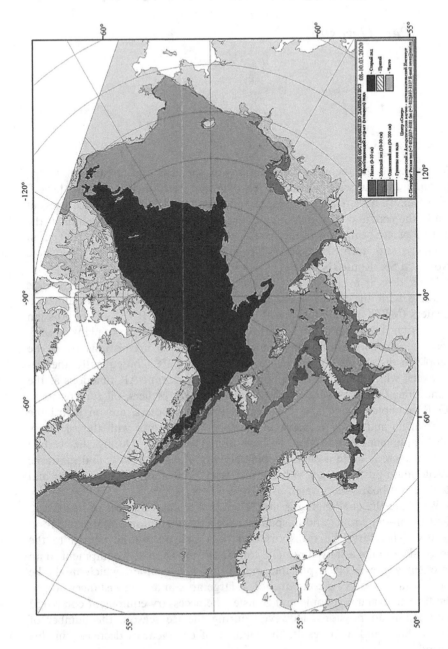

Figure 16.1 Ice formations in the Arctic

In recent years, a series of emergencies has happened to small-size fishing vessels, and the number of boats has risen sharply during the fishing season. A lack of supervisory control of pre-voyage preparedness for these vessels and a lack of emergency equipment on board have caused an increasing number of emergencies. The main reasons for the emergency calls are engine and communication failures. The small-size vessels operate mostly near a coast, and this area is risky, owing to high cliffs, shallows, and storms. Sometimes large maritime deep-draft tugs ready for SAR operations cannot approach a boat in distress near the coast. The solution to this problem could be the control of communication and emergency equipment on board the small-size vessels organized by supervisory bodies before sailing and deploying shallow-draft rescue ships in the areas of concern.

Different types of voyagers and scientific expeditions operating in the Arctic have become one of the biggest challenges in recent years. Detailed route-planning and communication technology preparations, co-ordinated with coastal rescue organizations, are of great importance before starting to explore the Arctic, especially if the route passes through the ice of the Arctic Ocean. If these basic rules are ignored, the expedition can lead to tragic consequences. In some incidents, the MRCC cannot provide communication and radio contact with the voyagers because they do not have appropriate communication means, or their communication equipment does not match the MRCC's technical capabilities.

The SAR operations are not similar to one another; they can be united only by name of classification – medical evacuation, fire, hull breach, vessel collision, grounding, etc. However, in practice every SAR scenario is different. It is very difficult to compare the emergency missions from the viewpoint of how the situation is unfolding on board the vessels, the psychological response of different people to emergencies, the kind of stress that prevents the distress vessel crew from making the right decision, and many other circumstances. The success of SAR co-ordination critically depends on the competences of the MRCC and MRSC staff directly working with the masters of the vessels in distress. For that reason, high demands are placed on the staff of the MRCCs. The best candidates are masters and chief mates with at least five years' experience of international shipping. This professional experience can help in advising the master of the vessel in distress to rescue the crew. Twice in five years, the co-ordinators of the MRCCs take higher training courses and courses to obtain the Global Maritime Distress and Safety System Certificate. However, in my opinion, this is not sufficient. Joint courses with SAR and OSR co-ordinators of the neighbouring countries are needed. The joint actions providing SAR and OSR require vast experience, and the joint activities clearly face obstacles. The annual joint full-scale and table-top exercises to practise these tasks are of great importance. However, mostly only the strategic level leaders usually participate in such international exercises. That is why the joint co-ordination training involving SAR and OSR co-ordinators at the operational level directly at the MRCCs in Russia or JRCCs in Norway could be a remarkable experience.

Note

1 www.aari.ru/odata/_d0015.php?mod=1

17 Perspectives on Future Research within Crisis and Emergency Management in the Arctic

Odd Jarl Borch and Natalia Andreassen

Introduction

This book has illuminated some of the challenges of emergency response in the Arctic. Knowledge in this context is in demand to make the crisis management system work. This book spans a broad set of research areas and disciplines illuminating the need for cross-disciplinary research to understand managerial challenges related to, among others, nature, politics, communities, institutions, organizations and groups, and culture. In this chapter, we reflect on the knowledge gaps within the research areas highlighted in this book.

New Knowledge on Activity and Risk Aspects in the Arctic

Within the operational context field, we have illuminated the need to look into the increased number of stakeholders and operational patterns in the different regions of the Arctic. Risk aspects should be highlighted according to different types of commercial and governmental activity, including risk related to military activity and the new players in the region. Treaties and agreements, regulations, and standard operational procedures are established and revised as a continuous process of institutionalization. The institutions are to provide a framework for reliable operations and efficient crisis response. The institutions are developed through political processes both within the nation states, bilaterally and through international fora such as the Arctic Council and International Maritime Organization (IMO). There is a need to follow these processes, look into their antecedents, and evaluate their effects.

In this book we have included the situational awareness perspective as a scientific platform to complex social and natural features (Endsley, 1995; Endsley and Jones, 2012), as well as perspectives from criminal law and international conventions (Guilfoyle, 2009, 2014) and multi-faceted geopolitical dimensions (Seker and Dalaklis, 2016; Dalaklis et al., 2018). These perspectives can help to illuminate the contextual issues that have risk implications in the Arctic. Understanding the present legal and political framework and the impact of potential conflicts of interests and operations, including from newcomers to the discipline, are important to safeguard operations in the region. Not only could tensions and conflicting interests both in the civilian as well as military sphere increase, but centrum-periphery imbalances may also be present. The interest of the nation states, the local communities, and indigenous people could be at stake related to commercial and environmental issues, as well as the right to local resources. Cross-sectional research, including political science, law, geography, and institutional theory can provide the necessary understanding of the risk areas and the diverging interests of the Arctic region in time and space. As an example, the recent development of fast-expanding tailor-made fleets of so-called expedition cruise vessels, including vessels with ice-breaker capabilities (Marchenko et al., 2015), are taking some governments by surprise. Some shipping companies have ambitions to roam the most distant regions, including the North Pole. The implications for rules and regulations to routes and the coverage of search and rescue (SAR) capacities in different countries should be highlighted.

Access to critical data is needed for navigation safety (Pedersen, 2019) and for research within risk aspects in the Arctic (Trbojevic, 2000; Abbassi et al., 2017; Gudmestad and Solberg, 2019). Such data could include adequate environmental data with future predictions of ice or data on natural resources in the areas previously not accessible. Historical cases, story-telling, the subjective data of experiences, and more quantitative research are needed to understand the long-term trends and pitfalls of any operations in the Arctic. Data from several countries and regions could provide the necessary platform for comparative studies and a more complete picture of the present situation and future trends. We need more knowledge to negotiate a more stringent framework of operations to increase safety and security and to provide guidelines for safe operations.

A lack of information is the key contributor to the unpredictability of the Arctic. Operational risk is a central question for all operators and governments in the most challenging environments. The risk concept, even though it is widely used, is not very precise. The most frequently used approach is to look into the probability that something should happen and link it up to the (negative) accumulated consequences in both the short and long term. The factors influencing the vulnerability of persons, nature, organizations, and society could also be included as central elements in the part of the equation. Risk can be discussed related to the specific type of activity and the type of negative incident that could appear. As both the probability and the consequences depend on the region where the incident could happen, the risk should also be linked to specific areas of operations. A very important question to be dealt with is the activity level and the probabilities of accidents. People might think that an increased activity level could increase the risk. However, if the historical number of accidents is next to zero, and the technology and knowledge level is continuously improved, an increase in the number of vessels might not necessarily lead to increased risk. Also, with more activity, the consequences could be reduced, owing to more units in the neighbourhood that can serve as a vessel of opportunity. In several activity areas there will be more experienced planners and more tailor-made and adapted equipment. Laws and regulations such as the Polar code can contribute towards reducing risk down to zero. Here, we are in need of more knowledge on risk assessment.

The cruise industry has been subject to much scepticism, owing to the focus on the ratio of the number of passengers to the SAR capacity of the government. Mass rescue capabilities will always be in demand. However, the lack of serious incidents can indicate that the total risk of cruise ships in the Arctic is rather low, even though the numbers are increasing. For the past 30 years there have been no large-scale accidents in the Norwegian sector of the Arctic even thoug,h this is the area with highest traffic density. Even though there have been fires and engine troubles, the cruise vessels have managed to solve the challenges on their own. We are still using the *Maxim Gorky* ice floe collision in 1989 as an example of what can go wrong. The crew managed to save the ship themselves, although with the help from the Norwegian coast guard.

Thus, the risk concept has to be used with care. We need more knowledge and tools for an adequate estimation of risk based on multiple criteria. With a lack of statistics, a more qualitative approach could be necessary. The risk assessment tools at hand can mostly represent a bottom-up approach (Baksh et al., 2018). This could address the specific operation and type of consequence (people, environment, political) with the specific type of unit, as well as include a 'fault tree' analysis of what could go wrong in the specific sailing route at a given time, and include sample maps with the probability of certain hazards such as ice or icing (Fu et al., 2018). The decision analyses and learning from failures can provide valuable insights even from major disasters in non-Arctic environments of high uncertainty (Labib, 2014). Risk has to be related to the type of accident in question. This calls for data, method, and theory triangulation with a broad range of data types and sources, including technical knowledge for the operating vessels, methods to collect the data, including documents, interviews, observations, questionnaires, and varying theoretical interpretations. The categorization of risk levels could require qualitative approximation to the statistical background, as well as gaining experiences from similar incidents in other regions about failures that could emerge. Expert panels can be a valuable tool to prepare for unexpected events. Expert judgements and assessments of the operational conditions can reduce the complexity of the operational task environment.

Further validation of the risk assessment tools for achieving more precise measures call for the accumulation of a broader range of quantitative and qualitative data based on systematic research. One also has to start a discussion on the less likely events, making scenarios for 'black swan' events, where the probability is extremely low, but the consequences could escalate into a major disaster (Marchenko et al., 2018). Nobody would expect that a very modern cruise vessel like *MV Viking Sky* with four engines and a very competent engine crew should have a total 'black out', owing to low lubricant levels close to the Norwegian west coast and drift very close to the rocky shore in only half an hour (Accident Investigation Board Norway, 2019). With 1,373 people on board and with the weather making a launch of life-boats impossible showed that black swan events do exist. Engine trouble including fires on passenger ships is happening quite frequently. We need to know to what extent back-up systems exist and how fast they can be mobilized. In the *MV Viking Sky* case above, the crew managed to get the engines running after half an hour, avoiding a grounding by just 100 metres. As the different forces involved are moving targets, frequently mobilizing research-based expert panels for providing realistic risk matrices both on the probability and the consequence side can be of significance.

New Knowledge on the Institutional Platforms for Risk Reduction and Emergency Response

As the maritime activity pattern in the Arctic is changing, the institutions in the region have to follow suit. This includes both the institutions providing the necessary regulations regarding maritime activity, cross-border co-operation, and the empowerment of the emergency response agencies. Thus, we need to provide a more rigorous understanding of the responsibilities of the different governments and institutions to be proactive in facing new challenges. More often, one could claim that the institutions and not least governments are reactive, not placing enough focus and resources in an area before we have a severe accident. As an example, the introduction of the Polar Code by the IMO took several decades. After the *MV Viking Sky* cruise ship incident on the Norwegian coast in 2019, the Norwegian government reacted by establishing a committee to discuss cruise ship regulations. The degree of readiness of each national state has to be mapped, and the response capacity has to be evaluated, including the support from other nations through the so-called Host Nation Support Guidelines. Close co-operation and partnerships across borders are necessary to provide the necessary domain awareness, exchange of experience and solutions, sharing of capacity, and development of new rules and regulations. Providing a solid research platform for the initiatives taken within, among other bodies, the Arctic Council working groups are important. Both the Arctic Council and the Arctic Coast Guard Forum (ACGF) should encourage more research and develop a research policy in their area of responsibility.

We are in need of new knowledge about the professional facilitation of cross-border co-operation, including authority-sharing and delegation in maritime police matters (Roach, 2004; Mawby, 2007). Co-operation, knowledge, and activity are interlinked and interdependent. To administer new areas of co-operation, we need more knowledge on the laws and regulations, the political issues related to bilateral and international co-operation, and the environment for trust (Elgsaas, 2019). Joint operations in research and themilitary and commercial fields as well as within the industry associations, the exchange of experiences and plans between coast guards all provide the necessary platform for further communicative action. This could be of particular importance in areas like sharing resources, police work, and joint civilian-military operations, where research can provide the necessary knowledge platform for new agreements, as well as amendments of existing multinational treaties in the Arctic.

Establishing and understanding networking among the most important agencies like the Arctic coast guards, the police, and special task forces are vital. More organizational knowledge on the dependency and co-ordination of loosely coupled networks is critical (Brass et al., 2004; Raab et al., 2013). We also need more knowledge on the network of networks integration. As an example, the Arctic Council working group on Emergency Prevention, Preparedness and Response provides an umbrella network of coast guards, SAR agencies, oil spill response organizations, and radiation authorities. The coast guards are possibly the most important agencies in the Arctic, as they serve a broad range of government roles, and are literally on the borders of the other Arctic countries continuously.

An important task of institutions like the Arctic Council and the ACGF is knowledge exchange and remedy to co-operation challenges (Pincus, 2015; Østhagen, 2015). These institutions should also facilitate more thorough comparative research across their member countries. How these networks are linked together and share knowledge, tasks, and functions in a challenging political context is important knowledge. The understanding of the relations between the political levels, the formal networks as institutional platforms, and the operational levels in day-to-day operations is of importance for the management at every level.

New Knowledge on Crisis and Emergency Management, Competence, and Training

Most emergency response operations are taking place in predictable environments and can be characterized as 'business as usual'. Mass rescue operations, larger oil spill response actions, and other large-scale responses such as anti-terrorism and radiation incidents can put a heavy strain on the response system. The upscaling that can take place calls for additional management capacity and extra co-ordination. An understanding of the quality of the physical and human resources in an Arctic context and not least the transport and logistics issues are important research areas. Within SAR operations, fast response and the evacuation of humans are crucial, with helicopters as the dominating transport form. Within pollution response, mobilization takes a longer time, with a large amount of physical equipment distributed by boat. The operations can go on for several months, with the need for a large number of qualified personnel. Limitations in the number of units adapted for Arctic conditions can hamper the operation. In this field, the mapping of capacities and competences on response time and technologies is needed. Innovations related to logistics tools and operational systems including the human-machine interplay in an Arctic context should be supported by thorough research and testing. There is a need to look closer into the team level and how people from different regions and professions function together in rough, cold conditions. An actual question is how to co-ordinate the survival of vulnerable and injured passengers from cruise ships, keeping them alive until support can come. This research problem calls for a focus on creativity and flexibility in decision-making within a broader range of the emergency response system, including the personnel at the units in distress.

More units involved with increased interdependence between them and uncertainty about the impact of the performed actions increase the complexity of operation (Bigley and Roberts, 2001; Wolbers et al., 2018). Weather and other contextual factors can create added turbulence and unpredictability. As a starting point, the upscaling process can be based on standard operating procedures. However, there could be conditions that call for improvisation as to new solutions of managing and organizing, incorporating untrained units into an ongoing operation, and a response strategy calling for revised routines or organizational structures. We are in need of research showing how managers can build upon the standard plan systems, develop them further, and add new solutions, not least during an ongoing operation. Moreover, all tasks should be done within a very short time span, through rapid decision-making processes (Faraj and Xiao, 2006). As there are few large-scale operations in each country, there is a need for comparative studies of systems for continuous improvement and creation of new best practice and large-scale exercises for testing. This includes research and development on aspects like organizational hierarchies, roles, and authority and inter-organizational relations.

In this field, there is an essential need to be familiar with neighbouring countries' systems in the region in order to understand the expected routines for co-operation and command. Increased knowledge of complex systems shows how important inter-organizational co-ordination and the roles of the incident commanders at different levels are when facilitating joint emergency response in complex operations (Borch and Batalden, 2014; Borch and Andreassen, 2015; Andreassen et al., 2018). We need to focus further on the range of managerial roles and how they deal with the lack of predictability regarding emergency cause–effect relations, resources, and capabilities.

The joint configuration of an ad hoc emergency organization includes multiple actors across many jurisdictions with diverging organizational design (Kapucu et al., 2010). Researchers should look more closely into the structuring mechanisms to co-ordinate multi-agency collaboration (Comfort and Kapucu, 2006; Okhuysen and Bechky, 2009; Sydnes and Sydnes, 2011). Standardized command and control concepts have been developed and institutionalized over many years, based on common best practices. However, they can have their limitations, and 'one size does not fit all'. New technology and new threats can call for system innovations. Among others, knowledge of up- and down-scaling of the emergency response value chain and the co-ordination of several closely inter-linked organizations is in demand (Moynihan, 2009). Inter-organizational learning and building relationships and trust between organizations through the range of different training schemes and collaborative exercises is important (Crossan et al., 1999; Roud and Gausdal, 2019).

Among the most challenging situations for the emergency response system are cases of violent acts and terrorism at sea. In this field we find the 'black swan' events with low probability and huge consequences. The mitigation of consequences includes several parallel action types such as mass evacuation, a SAR operations, large-scale paramedics and medevac efforts, fire-fighting and pollution response, and not least a hard response action with police and special forces to neutralize the threat. The on-scene coordination in cases like this is most challenging. At the operational and strategic levels coordinators may also have to take very difficult decisions concerning life and health. The reaction has be fast and improvised to reduce the number of causalities. The response may also have political implications. We are in need for continuous research building upon good intelligence cooperation as to the threat pictures. We have seen examples of terrorist groups in other regions that attack the soft targets with low protection levels (Greenberg et al., 2006). Thus, in the Arctic, we are also in need of more knowledge on counter-terrorism operations. Counter-terrorism research has to build upon only a few cases, making the range of relevant data a challenging task in the research process. There is a need to focus on how to tailor-make counterterrorism measures on different combined situations, joint efforts involving personnel on board, and special forces or logistics support from other countries. We need task forces adapted to the modus operandi of different categories of violent actors and terrorists. Even though much of the operational knowledge has to be kept secret, an increased research focus on passenger ship terrorism in particular can evoke both better security measures and also represent a deterrence factor. Research focus can show that the response system is prepared to prevent or mitigate all types of terrorist attacks and protect all sorts of targets without revealing operational secrets. The systematic evaluation of implemented security measures and their robustness towards new variations in threats are important. As an example, the hybrid threat picture can represent additional factors that have to be dealt with, such as GPS and satellite communication jamming, fake news creation, cyber-attacks, etc. This call for inter-disciplinary, cross-sectional, and cross-border comparative research. Research co-operation between civilian and military research institutions should be encouraged.

The context issues, complexity, and limited predictability of large-scale operations in the Arctic represent a competence transfer challenge. Building relevant research-based educational programmes is a continuous process within academia. The competence programmes for professionals also have to include training and exercises. The frequency of training and realism are both important. Using simulators can provide an opportunity for frequent training and also provides a controlled area of systematic data collection for further research, including optimal training and exercise schemes. Understanding learning processes in emergency management development is of special importance. As training and not least full-scale exercises are costly, companies and organizations struggle to maximize their learning outcomes and develop a sophisticated approach to collaboration exercises. It is important with evaluation schemes to look into the effect of learning on organizational changes and effectiveness in managing emergencies. This calls for longitudinal studies to analyze if inter-organizational learning also leads to increased collaboration effectiveness in real operations.

In this chapter we have looked into the importance of good management and relevant competence for crisis and emergency response in the Arctic. There is a need for increased international co-operation in this field, also creating relevant research programmes both nationally and related to international fora. Joint Arctic institutions like the Arctic Council and the ACGF can play an essential role in mobilizing the research community for the common good of an improved knowledge platform.

References

Abbassi, R., Khan, F. , Khakzad, N. Vietch, B., and S. Ehlers. (2017). Risk analysis of offshore transportation accident in Arctic waters. *The International Journal of Maritime Engineering*, (January).

Accident Investigation Board Norway. (2019). Interim Report 12 November 2019 On the Investigation into the Loss of Propulsion and Near Grounding of Viking Sky 23 March 2019.

Andreassen, N., Borch, O.J., Ikonen, E.S. (2018). Managerial Roles & Structuring Mechanisms within Arctic Maritime Emergency Response. *The Arctic Yearbook 2018*: 275–292. Retrieved from: http://hdl.handle.net/11250/2591156.

Baksh, A.-A., Abbassi, R., Garaniya, V., and Khan, F. (2018). Marine transportation risk assessment using Bayesian Network: Application to Arctic waters. *Ocean Engineering*, 159: 422–436.

Bigley, G. A. and Roberts, K.H. (2001). The incident command system: high reliability organizing for complex and volatile environments. *Academy of Management Journal*, 44(6): 1281–1299.

Borch, O.J. and Andreassen, N. (2015). Joint-Task Force Management in Cross-Border Emergency Response. Managerial Roles and Structuring Mechanisms in High Complexity-High Volatility Environments. In Weintrit, A. and Neumann, T.(eds.), *Information, Communication and Environment: Marine Navigation and Safety of Sea Transportation*. (Boca Raton, FL: CRC Press).

Borch, O.J. and Batalden, B. (2014) Business-process management in high-turbulence environments: the case of the offshore service vessel industry. *Maritime Policy & Management*, Vol. 42, Issue 5: 481–498.

Comfort, L.K. and Kapucu, N. (2006). Inter-organizational coordination in extreme events: The World Trade Center attacks, September 11, 2001. *Nat Hazards*, Vol. 39: 309–327.

Crossan, M.M., Lane, H.W., and White, R.E.J.A. (1999). *An organizational learning framework: From intuition to institution. The Academy of Management Review*, 24(3): 522–537.

Dalaklis, D., Baxevani, E., and Siousiouras, P. (2018a). The Future of Arctic Shipping Business and the Positive Influence of the International Code for Ships Operating in Polar Waters. *Journal of Ocean Technology*, 13(4): 76–94.

Brass, D., Galaskiewicz, G., Henrich, J., and Tsai, W. (2004). Taking stock of networks and organizations: a multilevel perspective. *Academy of Management Journal*, 47(6): 795–817.

Elgsaas, I.M. (2019). Arctic Counterterrorism: Can Arctic Cooperation Overcome its Most Divisive Challenge Yet? *The Polar Journal*, 9(1): 27–44.

Endsley, M. and Jones, D.G. (2012). *Designing for Situation Awareness: An Approach to User-Centered Design*. (London, New York, and Boca Raton, FL: CRC Press).

Endsley, M. (1995). Toward a Theory of Situation Awareness in Dynamic Systems. *Human Factors Journal*, 37(26).

Faraj, S. and Xiao, Y. (2006). Coordination in Fast-Response Organizations. *Management Science*, 52(8): 1155–1169.

Fu, S., Zhang, D., Montewka, J., Zio, E., and Yan, X. (2018). A quantitative approach for risk assessment of a ship stuck in ice in Arctic waters. *Safety Science*, 107: 145–154.

Greenberg, M.D., et al. (2006). Maritime Terrorism: Risk and Liability. RAND Centre for Terrorism Risk Management Policy. Retrieved from: www.rand.org/pubs/monographs/MG520.html.

Gudmestad, O.T. and Solberg, K.E. (2019). Findings from two Arctic search and rescue exercises north of Spitzbergen. *Polar Geography*, 42(3): 160–175.

Guilfoyle, D. (2009). *Shipping Interdiction and the Law of the Sea*. (Cambridge: Cambridge University Press).

Guilfoyle, D. (2014). Piracy and terrorism. In Skordas, A. and Koutrakos, P. (eds.). *The Law and Practice of Piracy at Sea: European and International Perspectives*. Oxford: Hart Publishing: 33–52.

Raab, J., Remco, S., Mannak, B.C.. (2013). Combining structure, governance, and context: a configurational approach to network effectiveness. *Journal of Public Administration Research and Theory*, 25(2): 479–511.

Kapucu, N., Arslan, T., and Demiroz, F. (2010). Collaborative emergency management and national emergency management network. *Disaster Prevention and Management: An International Journal*, 19(4): 452–468.

Labib, A. (2014). *Learning from Failures. Decision Analysis of Major Disasters*. (London: Elsevier Science).

Marchenko, N.A., Borch, O.J., Markov, S.V., and Andreassen, N. (2015). Maritime activity in the High North – the range of unwanted incidents and risk patterns. The 23rd International Conference on Port and Ocean Engineering under Arctic Conditions (POAC 2015), Trondheim.

Marchenko, N., Andreassen, N., Borch, O. J., Kuznetsova, S., Ingimundarson, V., and Jakobsen, U. (2018). Arctic Shipping and Risks: Emergency Categories and Response Capacities. *TransNav, International Journal on Marine Navigation and Safety of Sea Transportation*, 12(1): 107–114.

Mawby, R.I. (2007). Plural Policing in Europe. In H.I. Gundhus, P. Larsson, and T.-G. Myhrer. *Polisiær virksomhet: hva er det - hvem gjør det?*. Oslo: Politihøgskolen.

Moynihan, D.P. (2009). The Network Governance of Crisis Response: Case Studies of Incident Command Systems. *Journal of Public Administration Research and Theory*. Vol. 19, No. 4 (Oct.): 895–915.

Okhuysen, G.A. and Bechky, B.A. (2009). Coordination in Organizations: An Integrative Perspective. *Academy of Management Annals*, Vol. 3, No. 10: 463–502.

Østhagen, A. (2015). Coastguards in peril: a study of Arctic defence collaboration. *Defence Studies*, Vol. 15, Issue 2: 143–160.

Pedersen, T. (2019). Polar Research and the Secrets of the Arctic. *Arctic Review on Law and Politics*, Vol. 10: 103–129.

Pincus, R. (2015). Large-Scale Disaster Response in the Arctic: Are We Ready? Lessons from the Literature on Wicked Policy Problems. *The Arctic Yearbook 2015*. Retrieved from: https://core.ac.uk/download/pdf/50709191.pdf#page=235.

Roach, J. A. (2004). Initiatives to enhance maritime security at sea. *Marine Policy*, 28: 25.

Roud, E. and Gausdal, A.H. (2019). Trust and emergency management: Experiences from the Arctic Sea region. *Journal of Trust Research*. Retrieved from: https://doi.org/10.1080/21515581.2019.1649153.

Seker, B. and Dalaklis, D. (2016). Maritime energy security issues: the case of the Arctic. *Geopolitica Revista*, Vol. XVI, No. 66: 171–186.

Sydnes, M. and Sydnes, A.K. (2011). Oil spill emergency response in Norway: coordinating interorganizational complexity. *Polar Geography*, 34(4): 299–329. doi:10.1080/1088937X.2011.620721.

Trbojevic, V.M.C. and Carr, B.J. (2000). Risk based methodology for safety improvements in ports. *Journal of Hazardous Materials*, 71: 467–480.

Wolbers, J., Boersma, K., and Groenewegen, P. (2018). Introducing a Fragmentation Perspective on Coordination in Crisis Management. *Organization Studies*, 39(11): 1521–1546.

Index

Note: **Bold** page numbers refer to tables and *Italic* page numbers refer to figures

Printed in the United States
By Bookmasters